VISIONARY MANUFACTURING CHALLENGES FOR 2020

Committee on Visionary Manufacturing Challenges

Board on Manufacturing and Engineering Design

Commission on Engineering and Technical Systems

National Research Council

NATIONAL ACADEMY PRESS
Washington, D.C. 1998

NATIONAL ACADEMY PRESS • 2101 Constitution Avenue, N.W. • Washington, DC 20418

NOTICE: The project that is the subject of this report was approved by the Governing Board of the National Research Council, whose members are drawn from the councils of the National Academy of Sciences, the National Academy of Engineering, and the Institute of Medicine. The members of the committee responsible for the report were chosen for their special competencies and with regard for appropriate balance.

This report has been reviewed by a group other than the authors according to procedures approved by a Report Review Committee consisting of members of the National Academy of Sciences, the National Academy of Engineering, and the Institute of Medicine.

This study by the Board on Manufacturing and Engineering Design was conducted under Grant No. DMI-9626585 from the National Science Foundation. Any opinions, findings, conclusions, or recommendations expressed in this publication are those of the author(s) and do not necessarily reflect the view of the National Science Foundation.

Library of Congress Cataloging-in-Publication Data

Visionary manufacturing challenges for 2020 / Committee on Visionary Manufacturing Challenges, Board on Manufacturing and Engineering Design, Commission on Engineering and Technical Systems, National Research Council.
p. cm.
Includes bibliographical references and index.
ISBN 0-309-06182-2
1. Research, Industrial—United States—Planning. 2. Production management—Technological innovations—United States. 3. Concurrent engineering—United States. [] I. National Research Council (U.S.). Board on Manufacturing and Engineering Design. Committee on Visionary Manufacturing Challenges.
T176 .V57 1998
658.5′7—dc21

98-40274

Printed in the United States of America.

COMMITTEE ON VISIONARY MANUFACTURING CHALLENGES

JOHN G. BOLLINGER (chair), University of Wisconsin, Madison
DENNIS K. BENSON, Appropriate Solutions, Inc., Worthington, Ohio
NATHAN CLOUD, Cirrus Engineering, Wilmington, Delaware
GORDON FORWARD, TXI, Dallas, Texas
BARBARA M. FOSSUM, University of Texas, Austin
DONALD N. FREY, Northwestern University, Evanston, Illinois
DAVID F. HAGEN, Michigan Center for High Technology, Dearborn Heights
JAMES JORDAN, NGM Knowledge Systems, Cupertino, California
ANN MAJCHRZAK, University of Southern California, Los Angeles
EUGENE S. MEIERAN, Intel Corporation, Chandler, Arizona
DAVID MISKA, United Technologies Corporation, Hartford, Connecticut
LAWRENCE J. RHOADES, Extrude Hone Corporation, Irwin, Pennsylvania
EUGENE WONG, University of California, Berkeley

Board on Manufacturing and Engineering Design Staff

ROBERT RUSNAK, senior program officer
THOMAS E. MUNNS, associate director
AIDA C. NEEL, senior project assistant
BONNIE SCARBOROUGH, research associate
ROBERT E. SCHAFRIK, director (until November 1997)
RICHARD CHAIT, director (after February 1998)

Board on Manufacturing and Engineering Design Liaison

WILLIAM C. HANSON, Massachusetts Institute of Technology, Cambridge

Government Liaison

GEORGE HAZELRIGG, National Science Foundation, Arlington, Virginia
BRUCE M. KRAMER, National Science Foundation, Arlington, Virginia

BOARD ON MANUFACTURING AND ENGINEERING DESIGN

F. STAN SETTLES (chair), University of Southern California, Los Angeles
ERNEST R. BLOOD, Caterpillar, Inc., Mossville, Illinois
JOHN BOLLINGER, University of Wisconsin, Madison
JOHN CHIPMAN, University of Minnesota, Minneapolis
DOROTHY COMASSAR, GE Aircraft Engines, Cincinnati, Ohio
ROBERT A. DAVIS, The Boeing Company, Seattle, Washington
GARY L. DENMAN, GRC International, Inc., Vienna, Virginia
ROBERT EAGAN, Sandia National Laboratories, Albuquerque, New Mexico
MARGARET A. EASTWOOD, Motorola, Inc., Schaumburg, Illinois
WILLIAM C. HANSON, Massachusetts Institute of Technology, Cambridge
JAMIE C. HSU, General Motors Corporation, Warren, Michigan
CAROLYN W. MEYERS, North Carolina A&T State University, Greensboro
FREDERICK J. MICHEL, consultant, Alexandria, Virginia
PAUL S. PEERCY, SEMI/SEMATECH, Austin, Texas
FRIEDRICH B. PRINZ, Stanford University, Stanford, California
DANIEL P. SIEWIOREK, Carnegie-Mellon University, Pittsburgh, Pennsylvania
GORDON A. SMITH, Vanguard Research, Inc., Fairfax, Virginia
JOSEPH WIRTH, RayChem Corp. (retired), Los Altos, California

The National Academy of Sciences is a private, nonprofit, self-perpetuating society of distinguished scholars engaged in scientific and engineering research, dedicated to the furtherance of science and technology and to their use for the general welfare. Upon the authority of the charter granted to it by the Congress in 1863, the Academy has a mandate that requires it to advise the federal government on scientific and technical matters. Dr. Bruce Alberts is president of the National Academy of Sciences.

The National Academy of Engineering was established in 1964, under the charter of the National Academy of Sciences, as a parallel organization of outstanding engineers. It is autonomous in its administration and in the selection of its members, sharing with the National Academy of Sciences the responsibility for advising the federal government. The National Academy of Engineering also sponsors engineering programs aimed at meeting national needs, encourages education and research, and recognizes the superior achievements of engineers. Dr. William Wulf is president of the National Academy of Engineering.

The Institute of Medicine was established in 1970 by the National Academy of Sciences to secure the services of eminent members of appropriate professions in the examination of policy matters pertaining to the health of the public. The Institute acts under the responsibility given to the National Academy of Sciences by its congressional charter to be an advisor to the federal government and, upon its own initiative, to identify issues of medical care, research, and education. Dr. Kenneth I. Shine is president of the Institute of Medicine.

The National Research Council was organized by the National Academy of Sciences in 1916 to associate the broad community of science and technology with the Academy's purposes of furthering knowledge and advising the federal government. Functioning in accordance with general policies determined by the Academy, the Council has become the principal operating agency of both the National Academy of Sciences and the National Academy of Engineering in providing services to the government, the public, and the scientific and engineering communities. The Council is administered jointly by both Academies and the Institute of Medicine. Dr. Bruce Alberts and Dr. William Wulf are chairman and vice chairman, respectively, of the National Research Council.

Acknowledgments

The Committee on Visionary Manufacturing Challenges would like to thank the participants in the Workshop on Visionary Manufacturing Challenges and those who responded to the Delphi Survey, which were the principal data-gathering mechanisms for this study. The information and insight from these sources were invaluable to the committee.

Presenters at the workshop on Visionary Manufacturing Challenges included Philip Burgess of the Center for the New West; Edward Leamer of the University of California at Los Angeles; Paul Sheng of the University of California at Berkeley; Wilfried Sihn of Fraunhofer Institute for Manufacturing, Engineering and Automation; H.T. Goranson of Sirius Beta; Rick Dove of Paradigm Shift International; Brian Turner of Work and Technology Institute; Steven J. Bomba of Johnson Controls; Debra M. Amidon of Entovation International; Richard Altman of Communication Design; Mauro Walker of Motorola; and Richard Morley of Morley and Associates. The committee would like to thank these individuals for the time and thought that went into the workshop presentations. In addition, the committee appreciates the presentations and discussions provided by Bruce Gaber of the Naval Research Laboratory and Edward Lightfoot of the University of Wisconsin.

The committee would like to acknowledge the efforts of Bonnie Scarborough, National Research Council research associate, in conducting the Delphi survey and Brian Borys, consultant, in analyzing the first round survey results.

This report has been reviewed by individuals chosen for their diverse perspectives and technical expertise, in accordance with procedures approved by the

NRC's Report Review Committee. The purpose of this independent review is to provide candid and critical comments that will assist the authors and the NRC in making the published report as sound as possible and to ensure that the report meets institutional standards for objectivity, evidence, and responsiveness to the study charge. The content of the review comments and draft manuscript remain confidential to protect the integrity of the deliberative process. We wish to thank the following individuals for their participation in the review of this report: Robert P. Clagett, University of Rhode Island; Richard Kegg, Cincinnati Milicron; Michael McEvoy, Baxter International; Robert Pfahl, Motorola; A. Alan B. Pritsker, Pritsker Corporation; H. Donald Ratliff, Georgia Institute of Technology; Joseph P. Wirth, consultant; and Joel Yudken, AFL-CIO.

While the individuals listed above have provided many constructive comments and suggestions, responsibility for the final content of the report rests solely with the authoring committee and the NRC.

Finally, the panel gratefully acknowledges the support of the staff of the Board on Manufacturing and Engineering Design, including Robert Rusnak, study director (until March 1998), Thomas E. Munns, study director (after March 1998), and Aida C. Neel, senior project assistant.

Preface

Emerging economies, social and political transitions, and new ways of doing business are changing the world dramatically. These trends suggest that the competitive environment for manufacturing enterprises in 2020 will be significantly different than it is today. To be successful in this competitive climate, manufacturing enterprises of 2020 will require significantly improved capabilities. The attainment of these capabilities represents the challenge facing manufacturing.

The recent pace of technological advances could lead to complacency and the belief that technology will be available "on demand." Today's advances, however, were the result of exploratory enabling research performed years ago. If manufacturing is to have the technical capabilities it needs in 2020, the research that will provide the scientific basis for these capabilities must be initiated now. This report identifies areas for investments in research and development that will meet the needs of future manufacturers. Although the focus of this report is on technology, new technologies and new business practices will be inseparable.

The objectives of this study were (1) to create a vision of the competitive environment for manufacturing and the nature of the manufacturing enterprise in 2020, (2) to determine the major challenges for future manufacturing, and (3) to identify the key technologies for meeting these challenges. To perform this study, the National Research Council, through the Board on Manufacturing and Engineering Design, convened a committee of 13 individuals with expertise in manufacturing operations, management, and practices; manufacturing technology; education and training; social, behavioral, and political science; and technology forecasting. The committee included representatives from small, medium, and large companies in a variety of industries. The results of the committee's deliberations on a wide range of material are presented in this report.

The committee solicited information from a variety of manufacturing experts as a basis for its deliberations. The key to the success of this approach was selecting participants who are visionary leaders who could look beyond today, including national and international leaders in manufacturing and representatives of progressive manufacturing organizations.

The opinions of the selected experts were communicated to the committee in two ways: (1) a workshop that provided a forum for manufacturing experts and experts in fields likely to affect manufacturing in the future (e.g., economics, geopolitics, workforce, and education) and (2) an international Delphi survey that elicited creative ideas and helped the committee prioritize future industry needs. Recent forward-looking manufacturing studies—including the Next Generation Manufacturing Project report, industry-specific technology road maps, surveys, and futuristic views of manufacturing—were also reviewed.

This study has several unique features. First, the committee's findings are derived from the international Delphi survey, the workshop, and the committee's deliberations with the assistance of experts in specific areas. Second, the committee identified the fundamental challenges that must be overcome for the realization of the collective vision generated by the participants. Third, the study focuses on a broad, international view of manufacturing in 2020 rather than on a particular industrial sector.

In the context of this study, "manufacturing" is defined broadly to mean the processes and entities that create and support products for customers. Manufacturing encompasses the development, design, production, delivery, and support of products. In the course of this study, it became increasingly clear that the definition of manufacturing will become even broader in the future as new configurations for the manufacturing enterprise emerge and the distinctions between manufacturing and service industries become blurred.

The committee understands that it would be foolhardy to think that the future can be precisely defined. Nevertheless, the needs identified in this report, which have been validated by visionary leaders of today, will be broadly applicable to all future manufacturing.

Comments on this report can be sent by electronic mail to bmaed@nas.edu or by FAX to BMAED (202) 334-3718.

JOHN G. BOLLINGER, *chair*
Committee on Visionary
Manufacturing Challenges

Contents

Tables, Figures, and Boxes

TABLES

FIGURES

BOX

VISIONARY MANUFACTURING CHALLENGES FOR 2020

Executive Summary

Manufacturing has changed radically over the course of the last 20 years and rapid changes are certain to continue. The emergence of new manufacturing technologies, spurred by intense competition, will lead to dramatically new products and processes. New management and labor practices, organizational structures, and decision-making methods will also emerge as complements to new products and processes.

Manufacturing will remain one of the principal means by which wealth is created. It is essential that the United States be prepared to implement advanced manufacturing methods in a timely way. A critical step in preparing for the future will be the development of an underlying technical foundation through research by industry, academia, and government institutions, which must be guided by a clear vision of manufacturing in the next century and an understanding of the fundamental challenges that must be met to realize this vision. In this study, the committee identifies fundamental manufacturing challenges that can guide current investments in research.

The Committee on Visionary Manufacturing Challenges was established by the National Research Council's Board on Manufacturing and Engineering Design (1) to create a vision of the competitive environment for manufacturing and the nature of the manufacturing enterprise in 2020, (2) to determine the major challenges for manufacturing to achieve the vision, (3) to identify the key technologies for meeting these challenges, and (4) to recommend strategies for measuring progress. The year 2020 was chosen to encourage thinking about revolutionary changes, rather than evolutionary advances based on current capabilities. The committee's findings are summarized below.

VISION OF MANUFACTURING IN 2020

The committee developed an information gathering process based on two primary mechanisms:

- A workshop was held for participants (primarily from the United States) representing a broad range of manufacturing expertise. The workshop included presentations and discussions on future trends in economics, business practices, environmental concerns, and manufacturing issues. A summary of the workshop presentations and discussions is included in the report as Appendix A.
- An international Delphi survey of manufacturing experts (more than 40 percent outside the United States) was conducted. Summaries of the survey methodology and results are included in this report as Appendices B and C.

The committee found that the experts who participated in the workshop and survey had a very positive collective vision for manufacturing in 20 years. The most important technical, political, and economic forces for the development of manufacturing are listed below:

- The competitive climate, enhanced by communication and knowledge sharing, will require rapid responses to market forces.
- Sophisticated customers, many in newly developed countries, will demand products that are customized to meet their needs.
- The basis of competition will be creativity and innovation in all aspects of the manufacturing enterprise.
- The development of innovative process technologies will change both the scope and scale of manufacturing.
- Environmental protection will be essential as the global ecosystem is strained by growing populations and the emergence of new high-technology economies.
- Information and knowledge on all aspects of manufacturing enterprises and the marketplace will be instantly available in a form that can be effectively assimilated and used for decision making.
- The global distribution of highly competitive production resources, including skilled workforces, will be a critical factor in the organization of manufacturing enterprises.

Manufacturing enterprises in 2020 will bring new ideas and innovations to the marketplace rapidly and effectively. Individuals and teams will learn new skills rapidly because of advanced network-based learning, computer-based communication across extended enterprises, enhanced communications between people and machines, and improvements in the transaction and alliance infrastructure. Collaborative partnerships will be developed quickly by assembling the necessary resources from a highly distributed manufacturing capability in response to market opportunities and just as quickly dissolved when the opportunities dissipate.

Manufacturing in 2020 will continue to be a human enterprise that converts ideas for products into reality from raw and recycled materials. However, enterprise functions as we know them today (research and development, design engineering, manufacturing, marketing, and customer support) will be so highly integrated that they will function concurrently as virtually one entity that links customers to innovators of new products. The form and identity of companies will be radically changed to encompass virtual structures that will coalesce and vanish in response to a dynamic marketplace.

New corporate architectures for manufacturing enterprises, including materials enterprises that convert raw and recycled feedstocks into an array of finished and semifinished materials and product enterprises that convert the new materials into configured products, will emerge. Although production resources will be distributed globally, fewer materials enterprises and a greater number of regional or community-based product enterprises will be connected to local markets. The product enterprises may be part of larger corporations, but they will be located in and serve local markets and will operate autonomously.

Extremely small-scale process building blocks that allow for synthesizing or forming new material forms and products will emerge. Nanofabrication processes will evolve from laboratory curiosities to production processes. Biotechnology will lead to the creation of new biosynthetic and bioderived manufacturing processes with new and exciting applications on the shop floor of the twenty-first century.

MANUFACTURING CHALLENGES

The two-part Delphi survey on visionary manufacturing challenges was designed to forecast manufacturing challenges in 2020 and to elicit information on enabling technologies for research and development. An analysis of the first questionnaire identified major challenges and enabling technologies for manufacturing enterprises. The second questionnaire identified the highest priority challenges and research topics based on the prioritized enabling technologies. The results of the Delphi survey are included in Appendices B and C of this report.

Based on the results of the Delphi survey, the committee identified six "grand" challenges for manufacturers that represent gaps between current practices and the vision of manufacturing in 2020.

Grand Challenge 1. Achieve concurrency in all operations.

Grand Challenge 2. Integrate human and technical resources to enhance workforce performance and satisfaction.

Grand Challenge 3. "Instantaneously" transform information gathered from a vast array of diverse sources into useful knowledge for making effective decisions.

Grand Challenge 4. Reduce production waste and product environmental impact to "near zero."

Grand Challenge 5. Reconfigure manufacturing enterprises rapidly in response to changing needs and opportunities.

Grand Challenge 6. Develop innovative manufacturing processes and products with a focus on decreasing dimensional scale.

KEY TECHNOLOGIES TO MEET CHALLENGES

This report identifies the technology areas with the greatest potential for meeting the grand challenges. The committee's judgments are based on the following criteria:

- Was the technology identified as a high priority technology in the Delphi survey?
- Was the technology identified as a high priority technology at the workshop?
- Is this a primary technology for meeting one of the grand challenges?
- Does the technology have the potential to have a profound impact on manufacturing?
- Does the technology support more than one grand challenge?
- Does the technology represent a long-term opportunity (i.e., is the technology not readily attainable in the short term)?

After evaluating many ideas, the committee selected 10 strategic technology areas as the most important for meeting the grand challenges. These technology areas are listed below (not in order of priority):

- adaptable, integrated equipment, processes, and systems that can be readily reconfigured
- manufacturing processes that minimize waste and energy consumption
- innovative processes for designing and manufacturing new materials and components
- biotechnology for manufacturing
- system synthesis, modeling, and simulation for all manufacturing operations
- technologies to convert information into knowledge for effective decision making
- product and process design methods that address a broad range of product requirements
- enhanced human-machine interfaces
- new educational and training methods that enable the rapid assimilation of knowledge
- software for intelligent collaboration systems

RESEARCH RECOMMENDATIONS

The committee then identified research opportunities to support the development of the priority technology areas. The committee's general findings are listed below:

- Many of the areas for research are crosscutting areas, that is, they are applicable to several priority technologies. Adaptable and reconfigurable manufacturing systems, information and communication technologies, and modeling and simulation are especially important because they are key to manufacturing capabilities in many areas.
- Two important breakthrough technologies—submicron manufacturing and enterprise simulation and modeling—will accelerate progress in addressing the grand challenges.
- Substantial research is already under way outside of the manufacturing sector that could be focused on manufacturing applications.
- Progress toward the goals recommended in the Next Generation Manufacturing study on the needs of the next decade would provide some fundamental building blocks for meeting the longer-term grand challenges for 2020. These research areas include (1) analytical tools for modeling and assessment, (2) processes for capturing and using knowledge for manufacturing, and (3) intelligent processes and flexible manufacturing systems.
- Because manufacturing is inherently multidisciplinary and involves a complicated mix of people, systems, processes, and equipment, the most effective research will also be multidisciplinary and grounded in knowledge of manufacturing strategies, planning, and operations.

Based on the findings and general conclusions, the committee developed the following recommendations for a research and development program in the priority technology areas.

Recommendation. Establish an interdisciplinary research and development program that emphasizes multi-investigator consortia both within institutions and across institutional boundaries. Establish links between research communities in the important disciplines required to address the grand challenges, including all branches of engineering, mathematics, physics, chemistry, economics, management science, computer science, philosophy, biology, psychology, cognitive science, and anthropology.

Recommendation. Focus long-term manufacturing research on developing capabilities in the priority technology areas to meet the grand challenges.

Recommendation. Establish priorities for long-term research with an emphasis on crosscutting technologies, i.e., technologies that address more than one grand challenge. Adaptable and reconfigurable manufacturing systems, information and communication technologies, and modeling and simulation are three research areas that address several grand challenges.

Recommendation. Establish basic research focused on breakthrough technologies, including innovative submicron manufacturing processes and enterprise modeling and simulation. Focus basic research on the development of a scientific base for production processes and systems that will support new generations of innovative products.

Recommendation. Monitor the research and development on technologies that will have significant investment from outside the manufacturing sector and undertake research and development, as necessary, to adapt them for manufacturing applications. Some applicable technologies are listed below:

- information technology that can be adapted and incorporated into collaboration systems and models through manufacturing-specific research and development focused on improving methods for people to make decisions, individually and as part of a group
- core technologies, including materials science, energy conservation, and environmental protection technologies

Recommendation. Industry and government should focus interdisciplinary research and development on the priority technology areas. Some key considerations for the long-term are listed below:

- understanding the effect of human psychology and social sciences on decision-making processes in the design, planning, and operation of manufacturing processes
- managing and using information to make intelligent decisions among a vast array of alternatives
- adapting and reconfiguring manufacturing processes rapidly for the production of diverse, customized products
- adapting and reconfiguring manufacturing enterprises to enable the formation of complex alliances with other organizations
- developing concurrent engineering tools that facilitate cross-disciplinary and enterprise-wide involvement in the conceptualization, design, and production of products and services to reduce time-to-market and improve quality
- developing educational and training technologies based on learning theory and the cognitive and linguistic sciences to enhance interactive distance learning
- optimizing the use of human intelligence to complement the application and implementation of new technology
- understanding the effects of new technologies on the manufacturing workforce, work environment, and the surrounding community

- developing business and engineering tools that are transparent to differences in skills, education, status, language, and culture to bridge international and organizational boundaries

MEASURING PROGRESS

One of the key factors in meeting the grand challenges will be monitoring the progress of technology development. The committee believes a detailed research agenda and timetable based on the grand challenges and priority technology areas for manufacturing in 2020 should be developed. However, detailed research agendas or timetables were beyond the scope of this study. Research road maps that could be used to monitor progress toward realization of the vision of manufacturing in 2020 should be established in follow-up technology seminars with focus groups exploring the priority technologies and potential research areas. Rather than trying to anticipate the advancements for a twenty-year period, the committee recommends that general long-term goals be established in each technology area and that detailed road maps be established for five-year "windows of commitment." This approach, similar to the approach of the Defense Advanced Research Projects Agency, would provide a reasonable time frame for technology incubation, with yearly reviews to monitor progress. At the end of the five-year period, goals and programs would be re-examined for the next five-year period. This approach would allow research efforts to be adapted to revolutionary advances and for unfruitful research directions to be reconsidered.

1

The 2020 Vision

The objective of this study was to identify technical challenges and enabling technologies for manufacturers to remain productive and profitable in 2020. The initial task facing the committee was creating a vision of the competitive environment for manufacturing and the nature of the manufacturing enterprise in 2020. According to Adelson and Aroni, ". . . the future is determined by innumerable decisions and actions interacting in rich and . . . indescribable ways" (Adelson and Aroni, 1975). Although anticipating every interaction that determines even a single event is impossible, anticipating future problems is the key to developing solutions to these problems. Future opportunities may require that the present plans be changed. Envisioning the future is, therefore, the key to influencing the future.

METHODOLOGY

The Committee on Visionary Manufacturing Challenges was established by the National Research Council's Board on Manufacturing and Engineering Design to identify the major challenges that will face manufacturing enterprises in 2020 and the enabling technologies to address these challenges. In addition to reviewing forward-looking manufacturing studies and industry visions (e.g., NGM, 1997; ACS, 1996), the committee used the following methods to develop a vision of the future:

- A workshop was held for participants (primarily from the United States) with a broad range of manufacturing expertise. The workshop included presentations and discussions of future trends in economics, business

practices, environmental concerns, and manufacturing issues. A summary of the workshop presentations and discussions is included in the report as Appendix A.
- An international Delphi survey of manufacturing experts (more than 40 percent outside the United States) was conducted. Summaries of the survey methodology and results are presented in Appendices B and C.

Based on the information gathered, additional presentations by leaders in manufacturing, and the committee's deliberations, the committee determined the major manufacturing challenges for the year 2020 and the enabling technologies that would be needed to address these challenges.

THE CONTEXT FOR MANUFACTURING IN 2020

For the purposes of this study, "manufacturing" was defined in broad terms as the processes and entities required to create, develop, support, and deliver products. Many forces—social, political, and economic, as well as technological—will shape the manufacturing environment in 2020. The committee concluded that the first step to envisioning the future of manufacturing was to envision the future world in general. Mechanisms to develop such a vision were included in the workshop and the Delphi survey. During the workshop, presentations were made on economic, social, and business trends; in the Delphi survey, participants were asked first to describe their view of the manufacturing environment in 2020. Although current trends must be considered to predict the future, the committee felt that the vision for manufacturing should be based on a positive view of the future that would be worth working towards.

Major Forces for Change

The nature of manufacturing enterprises will evolve in response to changes in the technological, political, and economic climate. The committee believes that the following factors will be the most important to the development of manufacturing:

- The **competitive climate**, enhanced by communication and knowledge sharing, will require rapid responses to market forces.
- **Sophisticated customers**, many in newly developed countries, will demand products customized to meet their needs.
- The **basis of competition** will be creativity and innovation in all aspects of the manufacturing enterprise.
- The **development of innovative process technologies** will change both the scope and scale of manufacturing.
- **Environmental protection** will be essential as the global ecosystem is strained by growing populations and the emergence of new high-technology economies.

- **Information and knowledge** on all aspects of manufacturing enterprises and the marketplace will be instantly available in a form that can be used for decision making.
- The **global distribution** of highly competitive production resources, including skilled workforces, will be a critical factor in the organization of manufacturing enterprises.

These trends suggest that flexibility and responsiveness will be critical for manufacturing in 2020.

MANUFACTURING IN 2020

Manufacturing Enterprises

Customers will require that suppliers of goods and services maximize the value relationships among quality, service, and price. The goal of successful enterprises will be to find the optimum position in this "better-faster-cheaper" competitive triangle. A "we *can* have it all" attitude among consumers will force corporations to become extremely flexible and adaptable. As large numbers of consumers in newly developed countries gain economic power, this attitude will be prevalent worldwide.

The concept of manufacturing in 2020 will be broader than it is today. It will include software (the conversion of information, as well as materials, into useful products), biotechnology, some aspects of agribusiness, and many other production enterprises. The basis for competition will be creativity and innovation because (1) the manufacturing context will be broader and (2) social and organizational structures will be much more knowledge-based, dynamic, fluid, and globally distributed. Manufacturing enterprises will plan, create, and manage new products, processes, supply chain systems, and other business aspects of the enterprise (e.g., finance and marketing) concurrently.

The structure and identity of companies will radically change to encompass virtual structures that will coalesce and vanish in response to a dynamic marketplace. All activities that are not essential to implementing new ideas in marketable products will be eliminated. A readily available generic transaction and alliance infrastructure (e.g., equitable profit sharing and business processes for protecting intellectual property) will enable individuals and entrepreneurial teams to compete solely on the basis of skills and knowledge. These developments will require new corporate architectures for manufacturing enterprises:

- **materials enterprises** that can convert raw and recycled feed stocks into an array of finished and semifinished materials to meet the changing demands of product suppliers in a cost-effective way
- **product enterprises** that can convert the new materials into configured products

Although production resources will be distributed globally, fewer materials enterprises and more regional or community-based product enterprises will be linked to local markets. Product enterprises may be part of larger corporations, but they will be located in and serve local markets and will operate autonomously. Materials enterprises will initially merge to achieve economies of scale, but this will change as revolutionary materials processes (e.g., molecular nanotechnology) enable the local production of specialized materials.

Companies will be aggregations of people connected to each other by mutual trust and supported by an alliance and transaction infrastructure. Companies will be characterized by their ability to define an increasingly fluid "core competency" in a supply chain. Core competencies will be perceived as commodities that can be combined and recombined in response to market dynamics.

Team-like organizations will form around new product ideas and quickly assemble the necessary resources from a highly distributed manufacturing capability. All participants will share decision making, risks, and rewards. All functions of the enterprise will be comprised of highly integrated systems of human, material, and information processing capabilities that can be combined to transform ideas and materials into valuable products. All aspects of developing a manufacturing enterprise, including developing business and marketing strategies, research, and product innovation, will be concurrent.

Enterprise teams will interact continuously with each other and with computer-based system synthesis models to explore the complete range of configurations and resources to realize new products. As a result, robust system configurations will be relatively invulnerable to external changes and highly adaptable to changes in technology, the marketplace, and the business climate. Adaptable enterprises will be able to reconfigure quickly to accommodate change while continuing to be profitable.

New systems technology will enable innovative processes to focus not only on developing new products, but also on creating optimal enterprise configurations. Sweeping changes will be based on technologies that are currently unfeasible or impractical. The integrated enterprise system will be dynamic, continuously changing in response to new opportunities. Societal-based economic considerations will drive businesses to optimize the responsiveness, quality, and profitability of the system.

Workforce

The manufacturing workforce will be as diverse as the global economy. Interpersonal skills will be highly developed, cross-cultural barriers will be greatly reduced, and remaining differences will be valued for their contributions to innovative manufacturing. Individuals will have a sense of purpose and satisfaction and will be able to see clearly how their skills and intellectual capabilities add value to the enterprise. Information systems that enhance workers' access to, and ability to use, information will reduce the current gap between individual intellectual capabilities. A diverse workforce, operating on a more level playing field, will have a greater potential for creating new products synergistically.

In addition to the workforce's situational adaptability, people, information/data processing systems, and material processing systems will be closely integrated. Individual workers will learn not only through access to information, but also by being important elements of a highly integrated manufacturing system. As automation advances toward more "human-like" capabilities, workers will be freed to do what is uniquely human—create valuable new products and make bold and visionary business decisions. The synergistic output of human-machine interactions will be much greater than the sum of its parts.

Process Technology

The innovation that has driven the microelectronics industry toward smaller and smaller processing scales could provide a model for revolutionary advances in industrial processes and equipment in the future. As enabling technologies are developed, the trend toward small-scale production components will continue. Extremely small-scale process building blocks that allow for synthesizing or forming new material forms and products will emerge. Nanofabrication processes will evolve from laboratory curiosities to production processes. Molecular assembly of complex, precise functional structures will lead to the development of microdevices, such as sensors, computational elements, medical robots, and macroscopic devices constructed from fundamental building blocks. Biotechnology, combining biology and chemistry, will lead to the creation of new biosynthetic and bioderived manufacturing processes that will have new and exciting applications on the shop floor of the twenty-first century.

The focus on sustainable, low-waste production processes will intensify as the global ecosystem is strained by growing populations and the development of new high technology economies and as awareness and global economic forces increase the need for responsible environmental stewardship. Improved process controls, the recycling and reuse of process waste streams, and new synthetic pathways will result in near-zero discharge processes. Products will be designed to be recyclable and reusable or to exist benignly in the environment.

SUMMARY

Predicting the future is a difficult but important task. Only by speculating about the future will we be able to affect it. Based on the information obtained from a workshop, an international Delphi survey, and a review of the literature, the committee developed a positive vision of manufacturing in 2020. The "grand challenges" presented by this vision of the future are discussed in Chapter 2.

2

Grand Challenges for Manufacturing

The vision for 2020 and beyond described in Chapter 1 suggests considerable changes in the manufacturing enterprise. The social and political environment, the needs of the marketplace, and opportunities created by technological breakthroughs will drive these changes. Moving from the current status of manufacturing to manufacturing in 2020 will present major challenges, which the committee defines as "grand challenges" or fundamental goals, that would make realization of the vision possible. The six grand challenges are listed below:

- achieve concurrency in all operations
- integrate human and technical resources to enhance workforce performance and satisfaction
- instantaneously transform information gathered from a vast array of sources into useful knowledge for making effective decisions
- reduce production waste and product environmental impact to "near zero"
- reconfigure manufacturing enterprises rapidly in response to changing needs and opportunities
- develop innovative manufacturing processes and products with a focus on decreasing dimensional scale

In this chapter, the grand challenges are discussed and enabling technologies for each challenge are identified.

GRAND CHALLENGE 1: CONCURRENT MANUFACTURING

Grand Challenge number 1 is to *achieve concurrency in all operations*. In the context of this report, "concurrency" means that planning, development, and implementation will be done in parallel, rather than sequentially. The goal is for the conceptualization, design, and production of products and services to be as concurrent as possible to reduce time-to-market, encourage innovation, and improve quality. Concurrent manufacturing enterprises will consider product support, including delivery, servicing, and end-of-life disposition (recycling, reuse, or disposal), during the design and production phases. All aspects of manufacturing will be networked so that informed decisions concerning one activity can be made based on knowledge and experience from all aspects of the enterprise. Feedback during the lifetime of products and services will be continuous.

Concurrent manufacturing will revolutionize the ways people interact at all levels of an organization. "Teamwork" is the word used to describe these interactions, but it may not accurately describe the relationships of the future. Interactive computer networks will link workers in all aspects of the business. New social relationships and communication skills will be necessary, as well as a new corporate culture in which success will require not only expertise and experience, but also the ability to use knowledge quickly and effectively.

Concurrency will drastically shorten the time between the conception of a product and its realization. For example:

- Consumer products that now take six to nine months to reach the market will be delivered to customers within weeks of conceptualization.
- Large products that are combinations of mechanical structures and electronics that now take years to develop will be put into service within months.
- Microprocessor design will be reduced to a two-month cycle supported by flexible fabrication facilities that can produce new designs in a month.
- Composite and synthetic materials will be available almost immediately after their properties have been specified for product applications.

Many competitive pressures will force the reduction of time-to-market:

- Market opportunities will arise and disappear quickly.
- Lot sizes or batch sizes will be small as customers demand products and services tailored to meet their individual needs.
- Rapid changes in available technologies will cause rapid changes in products and reductions in production costs.
- Competitors from all parts of the world will enter and exit markets rapidly as opportunities emerge and fade.

Concurrency is a natural response to the corporate enterprises envisioned in Chapter 1, in which core competencies and knowledge of different segments of the extended enterprise will be dynamically combined to meet specific, narrowly defined market opportunities. Accurate estimates, optimization, and tracking of product costs and revenues will greatly reduce financial risks.

Concurrent manufacturing is a grand challenge that will require not only significant new technologies in communication and processes by which products are conceived and produced, but also a new definition of the social and cultural environment of manufacturing organizations. This will be particularly important for global, multidisciplinary, multicultural, and highly transient organizations.

Current Status

The recent introduction of methodologies for integrated product and process designs and of integrated product teams has reduced time-to-market significantly (e.g., for recent automobile models[1] and microprocessors). But even the most advanced collaborative design software cannot incorporate tacit knowledge, respond to changing markets or organizational structures, or accommodate multilingual or multicultural projects.

Manufacturing enterprises today are struggling just to exchange design data. Exchange standards for product data, such as STEP (Standard for the Exchange of Product-Model Data [www.nist.gov/sc4/www/stepdocs/htm]), are just beginning to be accepted. Products that have been completely specified in digital form include the Boeing 777 (Computing Canada, 1997; CAD/CAM Update, 1997). However, exchanges of design data have been limited by the lack of interoperable systems-level applications software. The development of exchange standards for process data has been hampered by difficulties in characterizing and integrating processes. Enterprise resource planning is being implemented to manage resources more effectively, but large companies have encountered significant difficulties in the integration of enterprise resource planning with their design functions.

The development of designs that treat the entire life cycle of products is now a subject for academic research, but little has been done to integrate processes and life cycle costs and management into overall designs. Although high performance computers may eventually have sufficient computational capacity for comprehensive integrated designs (if models and simulations could be expressed and presented adequately), the optimization of product and process life cycles is still a distant possibility.

Technological advances promise to reduce time-to-market, although barriers to implementing them must still be overcome. Rapid prototyping technologies have shortened product development times and improved the integration of product and process design; experimental facilities for near-net-shape processes are beginning to build small quantities of parts and products. More flexible machine tools and manufacturing cells have reduced some set-up times from hours to minutes, although most manufacturing is still done by inflexible machine tools in

[1]The new Crysler Concorde and Intrepid models took 31 months to develop and bring to market, an improvement of 8 months over the first LH models, which in turn had been brought to market quicker than previous models (Reuters, 1997).

fixed cells with inflexible controllers. Major product lines, such as automobiles, often require weeks of down time and large capital investments in new machines and retooling when new models are introduced.

Enabling Technologies

The technologies, processes, and systems that support the small, tentative steps toward concurrency being taken today are primitive compared to the requirements for the future. New technologies will have to support new organizational concepts that can enable geographically distributed work units, with multicultural and multidisciplinary participation, to work concurrently and to adapt and change rapidly.

The level of concurrency envisioned by the committee will require technological advances in four key areas: systems modeling capability; modular, adaptable design methodologies; adaptable manufacturing processes and equipment; and materials and processes.

Systems Modeling Capability

Systems models that can synthesize all aspects of a manufacturing enterprise will ensure that operational decisions contribute to a feasible, even optimal, solution. Modeling and simulation of an entire manufacturing enterprise will be used in concurrent, enterprise-wide planning and for making real-time operational decisions. Future systems models must incorporate all aspects of manufacturing, including equipment, processes, and the ways people interact with them in manufacturing systems (e.g., human-machine interfaces and processes and subsystems that enhance human performance and promote intelligent input). The issues are more complex than simple ergonomics and include considerations of human cognition and learning.

Modular and Adaptable Design Methodologies

To support concurrency, designs will have to be readily adaptable to a broad range of products, processes, and process parameters. Design methodologies will draw on libraries of reusable design modules that consider waste generation, raw material and resource utilization, manufacturing costs, maintenance time, and other parameters.

Adaptable Processes and Equipment

Concurrent manufacturing will require processes that can be rapidly adapted to manufacture new products to meet dynamic market demands. Producing several customized products on the same process line will require adaptable manufacturing processes and systems that can be quickly reconfigured. Digital representations of product designs will have to be developed quickly and transformed into finished products with minimal set-up time or human intervention. Process designs will have to flow seamlessly into machine or process set-up and product fabrication based on programmable, net-shape, flexible forming processes that do not require hard tooling. Modular equipment will be used whenever possible, with integratable, "plug-and-play" hardware and software components.

Materials and Processes

The rapid realization of new products will require processes that can produce totally new materials and shapes. These processes are likely to make use of new materials with new properties and structures. For example, large production-quality components with varying material properties and high dimensional precision can be produced using free-form fabrication. Materials for one-of-a-kind products may have to be created just for one use. Customizing new materials and shapes will require that processes be controllable at the atomic level to produce synthetic materials to meet specific, perhaps novel, performance objectives. This will necessitate the development of modeling capabilities that can derive the properties of the bulk materials from representations of atomic structures.

Most biotechnological manufacturing will involve systems that use biological processes to produce materials defined at the molecular level and then use these materials to produce finished products. Biotechnological manufacturing will also involve complex organic subprocesses, similar in some cases to processes used in the chemical industry; in other cases, biotechnology will involve organic growth.

GRAND CHALLENGE 2: INTEGRATION OF HUMAN AND TECHNICAL RESOURCES

Manufacturing technologies will continue to be planned, operated, maintained, coordinated, and enhanced by people in the year 2020. A global, competitive, fast-changing environment will make technology increasingly dependent on people. Technologies will have to be capable of adapting to the changing needs of the market, and people will have to know how to optimize and enhance them. Grand Challenge number 2 is to *integrate human and technical resources to enhance workforce performance and satisfaction.*

Manufacturers will be under tremendous competitive pressures to customize their products. Individuals and teams will have to be agile to maintain control over time and technology and capitalize on both. Successful organizations will have to educate their workers to consider time and technology as challenges to productivity, and workers at all levels will have to be knowledgeable about their products, the markets and customers that buy them, the processes used to make them, and the way their businesses operate. Whether manufacturing enterprises are part of a corporation or part of a network, they will have to be small, flexible, and highly competitive. Manufacturing enterprises will require integrated systems, automated routine functions, and people dedicated to finding solutions to address customer's needs.

Analysis of Chapter 1 shows that five principal factors will compel the integration of human and technical resources:

- To meet market demands, all members of the workforce will have to react quickly to customers, who will have high expectations and many choices.
- The rapid response environment will require effective communications at all levels of an organization, especially with customers, suppliers, and partners.
- The rapid assimilation of new technologies will require rapid learning throughout the enterprise.
- Frequent reconfigurations will require that enterprises adopt a systems approach.
- Successful enterprises will require that workers be self-motivated and have a sense of ownership of manufacturing and business processes.

Enterprises that can teach workers new skills quickly will have a competitive edge. Technologies that facilitate continuous learning will be essential. These technologies will be capable of making quick simulations of the likely consequences of future events and will allow people to acquire and use the results of the simulation quickly.

Manufacturing centers will operate within networks. Although the networks might be regional or community-based, they are likely to include other manufacturing centers around the world. The network will also include suppliers, partners, and customers. Highly skilled, knowledgeable workers will have to be able to communicate effectively within the enterprise, and direct communications between workers and customers will be commonplace. Workers will be able to communicate directly with customers based on their comprehensive understanding of the organization. Decisions to transfer critical work to a supplier or partner will be made by workers, who will be directly responsible for producing the product. In other words, those who are closest to the manufacturing process will be the ones who make promises to customers about product features, delivery, and price.

Because enterprises will have to be reconfigured frequently to meet production demands and new processes and products will be introduced continuously, job requirements will be constantly changing. To cope with these demands, each employee or employee group will have to become a business unit manager or a member of a business unit management team. Individual workers will continue to have specialized technical skills, but they will share their knowledge much more freely than they do today. Workers will have to make judgments that affect, and are affected by, the entire supply chain. In this environment, the business unit manager's responsibility will extend not just from "stock to dock," but also to strategic planning, market research, community outreach, product/process design, and recycling.

As increasingly complex technologies are developed, new ways will have to be developed to analyze and implement them in ways that workers can readily understand and use. Workers will have to be able to integrate technology into their daily work in ways that take advantage of the benefits of new technologies. Technologies will have to be configurable to the needs of individual workers. In addition, workers will have to be both skilled and experienced in many functions and disciplines of manufacturing to appreciate the enterprise as a whole.

Factory configurations will have to be less structured than they are today so that, in most situations, workers will be able to reorganize equipment and processes to meet customer demands. Detailed process design and planning will have to be accomplished by working teams, with minimal involvement from management. The workers will determine when and if automation will contribute to the speed and quality of production.

Maintaining worker enthusiasm and acceptance will be crucial in a world of highly mobile workers. Worker and employer loyalties will have to be replaced by new values and rules that will benefit both. Worker performance will have to be measured by a worker's ability to synthesize knowledge to make effective decisions in the face of uncertainty and the ability to motivate others. Outcome will be of paramount importance in this reward system. A worker's knowledge of technology and manufacturing and business processes may be the basis for judging their ability to contribute to the overall system. As a result, workers will have to strive to become more knowledgeable to enhance their decision-making capabilities and sense of ownership, which in turn will enhance their enthusiasm and motivation.

A worker's level of knowledge, enthusiasm, and motivation will make them valuable in the marketplace. Workers in this climate will need a wide range of skills, including strategic planning, market analysis, engineering design, supply chain management, finance, production planning, and order fulfillment. Although not everyone in the manufacturing enterprise will be expert in all skills, the more skills an individual has, the more valuable they will be to the organization. Workers will need a supportive work climate and technologies that support this continuous learning process. The fear associated with changing jobs, companies, or even moving to another region or country must be mitigated by the transferability of a "dossier of knowledge and experience."

Current Status

New manufacturing technologies must be implemented by people. Some isolated technologies today depend on the integration of human and technical resources, but they are few and far between. Moreover, new technologies can be difficult to implement and maintain, which can slow the rate of innovation.

The manufacturing technologies of today were not intended to support just-in-time user learning, knowledge creation, and flexible use. Most current user interfaces for manufacturing technologies are based on the concept that a profile of what a user needs to know now and in the future can be created. As a result, fixed-formatted interfaces focused on reducing user errors have been developed instead of flexible-formatted interfaces that would encourage the user's creativity. Object-oriented programming, which makes more flexibility feasible, is not yet widely used.

Most manufacturing organizations today have independent databases and tracking systems focused on specific functions or different stages in the supply chain. Although flexible manufacturing systems and computer-integrated manufacturing have been developed, enterprise integration is still a dream rather than a reality for most organizations.

Most effective collaboration today still takes place among a few partners in similar disciplines (e.g., engineers) across a narrow slice of the supply chain using a standardized software package for the interface (e.g., the use of CATIA by Boeing for the 777 [CAD/CAM Update, 1997]). Anyone not comfortable using the standard software package (e.g., small suppliers) may have difficulty adjusting to the collaboration technology. No collaborative tools are available today that would make it possible, for example, for operators and engineers to collaborate virtually unless both are working with the same standard engineering design package. Tools will have to be developed (e.g., CASE tools) to identify inconsistencies in language and create dictionaries that can be used by people in different disciplines.

Efforts to include operational personnel in planning and design activities have been largely unsuccessful so far. Difficulties in the transition to the skilled and empowered workforce envisioned for 2020 include the education and training of a more sophisticated and skilled workforce and the development of human-machine interfaces and enterprise configurations that can account for all skills and interests.

Current production process simulations are primitive and require that operators have specialized knowledge of process models and software tools to run them. Moreover, models of manufacturing operations are usually oversimplified. For example, models of factory processes, integrated with scheduling systems, are of limited use because they do not include human factors like variable skills, discretion, or motivation.

The manufacturing technology of today, including discrete-parts, batch, and continuous manufacturing, can only be reconfigured in very limited ways and only with significant human intervention. By the time data have been input, non-integrated systems have been coordinated, and error-filled programs have been fixed, the market may already have shifted.

Finally, manufacturing process technologies today often relegate people to unimportant or routine work. Tasks are allocated to automated processes first, and humans are assigned the leftover tasks. This means that (1) routine activities are assigned to workers if machines that can perform these tasks are too expensive, and (2) workers who are assigned to perform the more intellectual, non-routine tasks may be too distant from the production process to make effective decisions.

Enabling Technologies

The challenge for 2020 is to develop technologies that enhance people's intellectual contributions to their work, provide people with information and coordination capabilities for the total supply chain, and help people make informed decisions in the face of uncertainties. Manufacturing technologies of the future must perform the following functions:

- ensure that people are always learning when they perform a task
- provide people with real-time information on the status of each step in the supply chain, from market surveys through production to customer use
- enable people to collaborate seamlessly at any stage in the supply chain
- enable people to simulate alternative operational decisions in the face of uncertainties
- enable people to reconfigure processes and products rapidly to adjust to changing market needs without human involvement in routine operations
- provide people with the skills and knowledge to use their time for non-routine tasks, and leave routine labor to machines
- provide systems that effectively operate multicultural networks of people and machines

Substantial technological and sociological advances will be necessary for the development of optimal human/technical systems. The committee has identified the technical areas described below as the most important for developing manufacturing systems that can integrate human and technical resources.

Systems Models for all Manufacturing Operations

Systems models for all manufacturing operations will be required to facilitate operational decisions and dynamically allocate tasks to workers and machines. No single model to describe an entire manufacturing enterprise is available today, although models of various processes or operations of a manufacturing enterprise are available. Systems models for manufacturing operations will be needed to allow the user to apply any model in making decisions, test hypotheses with more than one model, and add knowledge and data to models to improve their utility. Most current models are missing significant information, such as information about worker motivation, competing organizations, and community interests and needs.

Technologies for Converting Information into Knowledge

Information and the ability of people to convert information into useful knowledge are core capabilities for integrating human and technical resources. Information technologies that enhance the synthesis of information and provide multiple views of process information, alternative interpretations, and guidelines for selecting among those views will be needed. An individual's ability to choose among uncertain alternatives will be facilitated by technology that can search for possible alternatives, present that information in a form suited to the individual's learning style, and help test alternative hypotheses in real time.

Unified Methods and Protocols for Exchanging Information

Standards for the exchange of information between people, between organizations, between people and machines, and between machines is fundamental to the integration of human and technical resources. Unified methods and standards of communication will be required to allow people to move from one process to another and modify that process easily as new information becomes available. Protocols for communications will provide significantly more capability, but also greater flexibility, than they do today to allow for intelligent human-machine interactions.

Processes for the Development, Transfer, and Utilization of Technology

Processes and incentives will have to be developed to keep people abreast of rapidly changing technologies, the information needed to apply their knowledge, and the means to disseminate the new information to others. New technologies that will allow people to assess the applicability of a new idea immediately, route it directly to those who need it most, and provide simulation capabilities to experiment with the idea are needed.

New Educational Methods

People will have to learn while they perform a task. Technologies that enable learning will provide people with models of causes and effects, ways to aggregate information to make optimal use of their strengths, and ways to experiment in a safe but realistic environment. Finally, educational technologies could help avoid "cognitive rigidity" by critically assessing stereotypical responses.

Design Methodologies That Include a Broad Range of Product Requirements

Future design methodologies must include a broader range of people in the design process to integrate human and technical resources effectively. Products can be flexibly produced only if the key stakeholders in the entire product life cycle are involved in their development. Although some companies now include suppliers and after-market customers as stakeholders, the list of stakeholders in the future will be much longer and will include process operators, process maintenance organizations, and product maintenance personnel. Including all of these stakeholders in product design will require resolving the enormous difficulties associated with variations in disciplines, knowledge, and languages. Thus, manufacturing enterprises will need technologies that enable them to design products graphically rather than digitally or to replace abstract performance criteria with functional and virtual prototyping.

Design Methods and Manufacturing Processes for Reconfiguring Products

Reconfiguration processes that require little or no human intervention would free people to become business unit managers. Future technologies should allow workers simply to set the new parameters for a product and inform them graphically of the characteristics, functional uses, limitations, and marketability of the product. A worker would then just "press a button" to have the product made.

New Software Design Methods

Software will no longer be designed by the waterfall method. Methods of participative design and contextual inquiry for designing software and information systems will be widely used. Technologies that support participative design and contextual inquiries, especially computer-based technologies, are required to accelerate the design process and enable the cross-site sharing of knowledge acquired during the design process.

Adaptable, Reconfigurable Manufacturing Processes and Systems

Adaptable, reconfigurable manufacturing processes will respond to the needs of individual workers. For example, a technology might be able to sense the condition of a worker and dynamically reallocate work. Technologies that can sense the condition of the customer, inform the worker, and suggest alternative ways to allocate work are examples of adaptable processes.

Human-Machine Interfaces

Human-machine interfaces must be optimized for people to perform dynamic, real-time scheduling, planning, maintenance, operation, and process improvements. Technologies that enable people to input and retrieve information verbally, graphically, and dynamically could significantly enhance their ability to use computers efficiently. An optimal system will help operators choose what information is used. The technology must enable rapid adjustment to changing situations and offer the user different kinds of information depending on the situation. The effect of an individual's naturalistic, ecological, and situational cognition on his or her ability to interpret information accurately and quickly will be critical, especially as businesses move toward more decentralized organizations. Finally, the format for presenting information will be a critical aspect of the interface. Technologies to sort through a voluminous amount of information, tailor the information to the user's changing needs, and determine the most readily understandable way to present the information will be required. Thus, the technology will not only customize formats for different users, but will also customize those formats for the urgency of the situation, the user's decision-making style in a given situation, and the nature and type of information being conveyed.

GRAND CHALLENGE 3: CONVERSION OF INFORMATION TO KNOWLEDGE

Manufacturers are already fundamentally dependent on information technology, and the dependency will increase in the future. Grand Challenge 3 is to *"instantaneously" transform information from a vast array of diverse sources into useful knowledge and effective decisions.*

The final report of the next-generation manufacturing study (NGM, 1997) suggests that manufacturers will have to be distributed worldwide to meet customer demands economically. This globalization implies the decentralization of the workforce, which will increase the need for fast, accurate, high quality communications. Because globalization also entails crossing national boundaries, communications will have to be transparent to language and cultural differences. Workers will have to be trained quickly, often at a great distance from the sources of knowledge and expertise. Networks of companies and alliances will have to be created and dissolved in response to rapid changes in business conditions.

Manufacturing enterprises are fundamentally and inescapably dependent on information technology, including the collection, storage, analysis, distribution, and application of information. If the exponential growth of computer and communication technologies (hardware and software) continues at its present rate, businesses of 2020 should be up to the task. The two main challenges will be (1) to capture and store data and information "instantaneously" and transform them into useful knowledge and (2) to make this knowledge available to users (human and machine) "instantaneously" wherever and whenever it is needed in a familiar language and form.

One of the challenges for future manufacturing will be to reduce the lead time for new products. Concurrent design and manufacturing will require the real-time transfer of information between designers and manufacturers. The global distribution of manufacturing resources and expanded supply networks will challenge information systems to maintain operating control. The logistics of physically moving materiel and inventories will also require real-time transaction-based information systems.

Perhaps the biggest challenge will be in education. Well trained, educated people will make better and faster decisions based on an unprecedented flow of data, information, and knowledge. Only trained and educated people will be able to separate useful information from useless information.

Current Status

A significant portion of U.S. manufacturing is done by companies with fewer than 100 employees. These small companies, which make up a large portion of the supply chains for large public companies, are often undercapitalized and are not usually on the cutting edge of technology. Small companies have limited access to inexpensive, easy to use information systems linked to the information systems of large companies.

Information technology is often adapted for manufacturing operations by people who are knowledgeable in information technology but not business operations. Consequently, investments in information technology in manufacturing environments have not resulted in the anticipated increases in productivity. It has been estimated that 50 percent of all new business information systems projects fail to attain their economic or operational objectives. The economic effects of information technology on sales, the cost of goods, capital returns, and other economic metrics are not well understood.

Many manufacturers feel that the current system of education—including primary and secondary education, vocational training, and undergraduate education—does not prepare employees for high-technology jobs. For example, universities have been increasingly challenged to train students in the use of advanced information technology to address business basics, including financial analysis and human factors. A truly interdisciplinary curriculum that considers information technology in a global context would make information systems much more useful to manufacturing enterprises.

Enabling Technologies

Educational Technology

The technologies used for education and training will have to change to meet the needs of the workforce as more and more enterprises become global, required job skills dramatically change, new technology and manufacturing processes are introduced, and the mobility of the workforce increases. Computer-based training will become the norm. The dynamics of teaching people quickly and remotely will impose significant challenges on instructors, students, and information and communications technologies. A major task will be to create tools independent of language and culture that can be instantly used by anyone, regardless of location or national origin.

Collaboration Technology, Teleconferencing, Telecontrol, Telepresence

As jobs and factories are distributed around the globe, real-time information technology will be the most effective means of collaboration. Tools will have to be developed that allow for effective remote interaction. Collaboration technologies will require models of the dynamics of human interactions that can simulate behaviors, characteristics, and appearances to simulate physical presence. Behavioral and social scientists who can ease the transition to virtual space will be essential members of development teams.

Natural Language Processing

Advances in education and collaboration technology will require communication tools with instantaneous translation capabilities that can go from language to language, even dialect to dialect, in written and oral communications. In some ways this technology could be considered an extrapolation of current trends, but implementation on a global scale will be difficult and complex and will require major technological advances.

Data and Information Filters and Agents

The sheer volume of information—including disinformation, garbage information, redundant information, wrong information, and useless information—will make data searching, filtering, and archiving indispensable. Intelligent agents, active knowledge filters tailored to individuals, and knowledge structuring tools will be needed to prevent "information overload."

System Security

Manufacturing enterprises of the future will be dependent on complete and accurate information. They will need efficient and foolproof security systems to protect data, information, and knowledge, which will be the lifeblood of the industrial enterprise, from theft, acts of malevolence, accidents, misuse, and ignorance. A loss of security could have catastrophic consequences. The greater the volume of information and data, the greater the challenge will be to protect it.

Artificial Intelligence and Decision-Making Systems

Artificial intelligence and decision support systems will manage the selection of data and information, as well as system security. Artificial intelligence, including expert systems, object oriented technology, intelligent agents, multimedia systems, voice recognition systems, and neural nets, has already made significant inroads into manufacturing technology and has the potential to make continued advances in the very near term. But the challenges of the future will make these real successes seem insignificant. Future systems will have to handle huge image bases in a variety of languages where small nuances could make big differences and where even small differences could become catastrophic.

Automatic Sensors and Actuators for Process and Equipment Control

As manufacturing enterprises become more and more automated, processes and equipment will have to be tightly controlled to ensure high quality, low cost output with minimum waste. Manufacturing enterprises will rely more and more on automatic and multifunctional sensors and intelligent controls on the process and enterprise levels (NRC, 1998). The design, manufacture, optimization, and effective deployment of these systems will be critical to process industries in the next century.

Integrated Modeling and Simulation

Validated and integrated enterprise models based on up-to-date information from distributed databases will enable people at all levels of an enterprise to make better and faster decisions. Models applicable to all levels of the manufacturing hierarchy, including equipment and process design, operations, distribution, service, and logistics, will be dynamically linked so the ripple effects of decisions will be available to other decision makers.

Intelligent models will mean significant savings of time and resources for manufacturing enterprises. These models will require improvements in computer and communication technologies, including visualization technology, computational speed, communications speed, and user interfaces.

GRAND CHALLENGE 4: ENVIRONMENTAL COMPATIBILITY

Grand Challenge 4 is to *reduce production waste and product environmental impact to "near zero."* The goal of manufacturing enterprises will be to develop cost-effective, competitive products and processes that do not harm the environment, use as much recycled material for feedstock as possible, and create no significant waste, in terms of energy, material, or human resources. Access to, and a working knowledge of, the global database on environmentally harmful materials will be a key element in meeting this challenge.

The world population has been projected to grow from 5.6 billion today to 8 billion in 2020 (NRC, 1996). The global ecosystem will be severely strained by this growth in population and the continued development of regions that currently have relatively low-technology economies, threatening the availability of resources and increasing waste.

The changes in environmental technology and environmental goals listed below were identified in a recent report by the National Research Council (NRC, 1996):

- the prevalence of incentive-based approaches to environmental regulation (instead of the command-and-control approaches used today)
- improvements in the measurement and monitoring of environmental quality to increase the understanding of ecological systems
- the reduction of adverse effects from chemicals in the environment
- the development of options for, and an assessment of the environmental impacts of, alternative energy sources
- the utilization of systems engineering and ecological approaches to reduce resource use
- a better understanding of the relationship between population and consumption as a means of reducing the environmental impact of population growth
- the establishment of environmental goals based on rates and direction of change rather than on specific targets

Future manufacturing industries will have a competitive advantage if they participate proactively in the assessment of environmental impacts, the establishment of environmental goals, and the development of technology to meet environmental goals.

Current Status

National, regional, and local governments are establishing standards that approach "near zero" pollutants in the environment. Manufacturing enterprises currently take one of three principal approaches to environmental management: remediation, compliance, or industrial ecology (Sheng and Allenby, 1997). Remediation is a command-and-control approach that involves treating wastes already in the environment to lessen their adverse effects. Compliance is also a command-and-control approach that involves government agencies establishing environmental standards for industry. Once industry has complied with a standard, the government often "raises the bar." Industrial ecology, or designing for the environment, is a strategic approach that involves preventing and minimizing environmental impacts over the entire product life cycle, from resource extraction to disposal (including recycling and reuse). More and more manufacturing industries are choosing industrial ecology as their approach of choice.

Current thinking about environmental compatibility is being driven by several trends, including emerging standards for managing product life cycles (e.g., ISO 14000), growing customer demand for "green" products, product take-back initiatives, and the internalization of all of the costs of waste disposal and abatement (Sheng and Allenby, 1997).

Consumers are becoming more aware of the environmental effects of the products they buy. In some cases, governments provide incentives for making environmentally conscious choices. Responsible environmental stewardship is becoming an increasingly astute business decision. Manufacturing enterprises with an environmentally friendly attitude have a competitive advantage in the more efficient use of resources through the recovery and reuse of process waste, increased use of recycled feedstocks, and more efficient processes that minimize waste generation.

Enabling Technologies

Manufacturers can identify and develop process technologies that will dramatically improve their use of energy, human resources, and materials. Manufacturing enterprises will face two principal environmental challenges. The first challenge is closing the gap between the current understanding of environmental impacts and technologies intended to reduce waste and control pollution and the understanding needed to meet future environmental goals. The second challenge is changing the spirit of the manufacturing enterprise to incorporate cooperation, proactivity, teamwork, and global partnering with governments, academia, allied and competitive manufacturing enterprises, and communities to reach environmental goals.

Modeling and Risk Assessment

A key challenge to manufacturing firms and to environmental regulators will be to provide a reliable base of environmental knowledge. The knowledge base will include accurate models of the effects of processes and materials on long-term environmental quality, quantification and comparisons of risks to the environment, and cost/benefit analyses that evaluate environmental choices or regulatory actions. The goal will be to create an inventory of environmental design criteria that includes assessments of impact that are universally accepted. The technology and credibility of environmental assessments will have to be greatly improved in terms of accuracy and credibility before regulators and manufacturers can establish common environmental goals.

Manufacturing Processes with Near-Zero Waste

Waste-free manufacturing will require design methods that consider the total life cycle of a product. Environmentally conscious manufacturers will evaluate waste production and recycling in each step of the conversion process, considering all process waste and by-products as "raw material" for other processes.

Environmental management will take advantage of advances in distributed information technologies. Innovative process technologies, such as net-shape processing, bioprocessing, and molecular self-assembly, could produce products with unique properties and characteristics and generate very little waste.

Reduced Energy Consumption

Processes optimized for near-zero waste often also require less energy. Ideally, the efficiency of mechanical energy and process heat will be maximized and the lost energy recycled, converted, or transferred to supplement the energy requirement. Particular attention should be directed toward recovering the immense amount of heat energy lost from metal processing furnaces, welding processes, coolants, transformers, compressors, condensers, and distillation columns.

Environmentally Aware Manufacturing Enterprises

A proactive approach to environmental compatibility will require changes in the "ethical spirit" of manufacturing enterprises. Industry regulators today tend to mistrust industry, particularly large enterprises. For the past 30 years, enterprises have been faced with a myriad of proposed, and enacted, regulatory environmental standards that directly affect people, materials, energy usage, and manufacturing processes. As a result, many enterprises now operate in a reactive and confrontational mode. Industry's perception that regulations are not based on compelling scientific analysis has made many enterprises reluctant to cooperate. Because of broadening responsibility of manufacturing enterprises for the global environmental impact of the products they produce, competitive enterprises in 2020 will have to cooperate closely with international environmental policy makers. Environmental goals and new process technologies that minimize deleterious environmental effects over the entire product life cycle must be developed with the consensus of all stakeholders and based on robust materials models and databases, assessments of environmental impact, and comprehensive risk assessments and cost/benefit analyses. A cooperative, collaborative atmosphere would encourage proactive industrial participation and the development of environmentally compatible processes.

GRAND CHALLENGE 5: RECONFIGURABLE ENTERPRISES

A significant challenge in the year 2020 will be the ability of an organization to form complex alliances with other organizations very rapidly. Grand Challenge 5 is to *reconfigure manufacturing enterprises rapidly in response to changing needs and opportunities.* Reconfiguration could involve multiple organizations, a single organization, or the production/process floor of a single organization. The driving factors for reconfigurable enterprises are rapidly changing customer needs; rapidly changing market opportunities; and developments in process, product, and electronic communications technology.

The ability of individuals and organizations to form complex collaborative alliances with other organizations will be a significant challenge. These relationships will have to be established and dissolved quickly to meet the challenges of increased access to (and demands of) less developed economies, rapidly changing markets, and expected advances in electronic communications. The challenge will be intensified by organizations having to cooperate and compete, simultaneously, with their "alliance" partners. Organizations will not be able to change their core competencies fast enough to take advantage of meaningful opportunities, so they will have to form alliances. Even multinational corporations will have to enter into alliances to take advantage of global opportunities. Rapid reconfiguration at the level of a single organization will require new organizational structures and employee relationships, as well as much greater flexibility and integration of activities.

Enterprises in 2020 and beyond will be characterized by capabilities and practices in the following areas:

- intraorganizational and interorganizational structures based on flexible, transient cooperation models
- enterprises focused on market opportunities rather than self-preservation and growth
- sharing of information and technology among competitors
- resolution of issues related to worldwide patents and other intellectual property rights
- equitable sharing of the rewards of collaboration
- incorporation of activity-based or knowledge-based values into transactions
- value-based relationships and value-based cost estimates
- well integrated, seamless supply chains
- cross-cultural systems of information management, representation, and communication

Current Status

Many exemplary current partnerships and alliances can attest to the challenge of developing long-lasting relationships based on trust and mutual benefit. Alliances today are not formed and dissolved quickly, although many organizations have successfully addressed the challenges posed by multiple cultures, different organizational technologies, different strategic priorities, different organizational structures and processes, and long histories of competition with one another. However, these barriers were not overcome quickly.

Increasingly, enterprises are adopting methods of measuring performance that account for intangible enterprise goals. One method, the "balanced scorecard" (Kaplan and Norton, 1996), supplements financial measures, such as return-on-capital and economic value added, with metrics that measure value to the customer, enhanced internal business processes, and employee learning and growth. A number of organizations have been able to restructure themselves in nontraditional ways and have made significant changes in their reward, performance management, and information systems to increase their flexibility and responsiveness. These changes have often been made at enormous expense.

Enabling Technologies

Enterprise Reconfiguration

Enabling technologies for reconfiguring enterprises include legal instruments, such as model agreements and contracts; models of qualitative socioeconomic factors; organizational and workforce relationships; and information technologies (computer applications and communications).

Software technologies can be grouped in an integrated software platform to support a common plan from conception to operation and include: standard terms to describe alliances; enterprise-wide system modeling; simulation modules that encompass legal, qualitative socioeconomic factors, and organizational and workforce relationships; and tools for theoretical analyses. The integrated software platform will guide the alliance through various stages from conception to fruition and provide both a shared medium for planning and a means for each participant to simulate the potential effects of decisions. Communications technologies will include uniform standards for exchanging manufacturing information, simple mechanisms for teleconferencing, and network protocols specific to the needs of the manufacturing alliance.

Organizational Reconfiguration

Forming and dissolving teams within a single organization is not very different from forming and dissolving alliances and will result in similar problems. Organizations would benefit from team theory (participants share common goals) more than game theory (participants have different and possibly conflicting objectives). Teams within a single organization will tend to focus on a single product or family of products and will require modeling, design, and simulation capabilities.

Reconfiguration of Manufacturing Operations

The production of diverse, customized products will require the rapid reconfiguration of manufacturing operations, with the following capabilities:

- systems models for all operations
- fundamental understanding of manufacturing processes
- synthesis and architecture technologies for converting information into knowledge
- unified communication methods and protocols for the exchange of information
- machine/user interfaces that enhance human performance
- adaptable and reconfigurable manufacturing processes and systems (e.g., biosynthetic processes and net-shape, programmable, flexible forming processes that do not require hard tooling)
- sensor technology for precision process control

Reconfiguration of manufacturing operations will involve different concepts and technologies than reconfiguration of enterprises or organizations. Here, the goal is to enable adaptation of manufacturing operations to make quick changes in the product or even to make different products. The realization of a reconfigurable factory requires tools that can combine basic operations in flexible ways to produce a set of processes, similar to the way linguistic primitives are combined using a flexible syntax to produce a rich variety of programs in a language. The degree of reconfigurability that can be achieved in this way far exceeds the degree of reconfigurability achieved by rearranging equipment.

The linguistic paradigm is well supported by a number of technology areas that are currently under active development. These include rapid prototyping tools, for both software and hardware; net-shape, programmable, flexible forming processes; cluster tools; and science-based process modeling. Software technologies, such as language design and compiler optimization, object-oriented and distributed databases, and virtual reality, would also yield great benefits if they were focused on the manufacturing context.

GRAND CHALLENGE 6: INNOVATIVE PROCESSES

The most significant advances in manufacturing in the past 25 years have been largely driven by information technology, computer tools, automation, and advanced work practices. However, the unit processes that transform materials into products have advanced only incrementally. Advances in the control of processes and microstructures at submicron scales and the analysis and unlocking of the chemical and biological secrets of nature will have an overwhelming effect on the future understanding of processes and chemical makeup. This will lead to new and exciting ways to manufacture, clone, grow, and fabricate a vast array of products. Grand Challenge 6 is to *develop innovative manufacturing processes and products with a focus on decreasing dimensional scale.* The challenge is to apply totally new concepts to manufacturing unit operations that will lead to dramatic changes in production capabilities. Significant advancements will be made possible by designing and processing products at smaller and smaller scales, ultimately at molecular and atomic levels. The need for these revolutionary processes will be driven by the competitive realities in 2020, when the primary differences between manufacturing enterprises will be their ability to create and produce new products rapidly to meet the high expectations and constantly changing demands of customers.

In the world of 2020, revolutionary unit operations will lead to dramatic new capabilities in the following ways:

- The integration of multiple unit processes into a single operation will significantly reduce capital investment, inspection time, handling, and processing time (NRC, 1992). Theoretically, a single machine could produce an entire product.
- Processes that are completely programmable and do not require hard tooling will enable the customization of products and rapid switching from one product to another.
- The creation of self-directed processes will simplify tooling and programming requirements and provide greater operational flexibility.
- Manipulation at the molecular or atomic level will lead to the creation of new materials, eliminate separate joining and assembly operations, and allow material composition to be varied throughout a single part.

Development of these innovative processes would enable the manufacture of new products, such as biological computers with molecular-sized components, molecular-sized surgical tools that could operate at the molecular or cellular level, efficient and inexpensive solar energy collectors, and new materials with significantly improved and tailorable properties.

Current Status

The underlying principles of current unit manufacturing processes have not changed in the past 25 years, which has limited advances in manufacturing capabilities. Mechanical or structural parts and products still require that processes be partitioned by function—material formulation, shaping, joining, assembly, and finishing. For the most part, processes across these functional categories are not integrated. Generally, each operation requires hard tooling and consequently has limited flexibility. Processes also vary widely in scale. The finest level in routine production is about 25 mm. In electronics, significant improvements in product performance have been made as a result of the ability to manufacture chips with higher and higher densities. Although the advances have been staggering, with current feature sizes at 250 nm (0.25 mm), they have been evolutionary. The basic processes have not changed fundamentally—mask production, deposition, and etching. The major focus of technology development today is concerned with the radiation used in lithography to allow finer and finer features.

Enabling Technologies

By 2020, revolutionary processes and capabilities will be based on technologies that are still in their infancy. One promising technology is direct materials deposition similar to the processes used for rapid prototyping. Although great strides are being made, this technology is used only for prototypes, not production parts. Using direct deposition processes for production parts will require two major breakthroughs: (1) the ability to use the processes with materials used in production parts, and (2) a significant increase in process speed.

Two technologies that could lead to the development of revolutionary processes are nanotechnology and biotechnology. Both of these will require major breakthroughs before they will be practical for manufacturing.

Nanotechnology

Nanotechnology would enable the development of new structures based on the precise control of materials architecture at the molecular or atomic level, tailored or functionally gradient structures with unique properties, and efficient and environmentally friendly processing. Molecular assembly methods will be required to enable nanoscale organization (NRC, 1994).

Nanofabrication technology includes the following types of processes:

- nanomachining (in the 0.1 to 100 nm range) to create nanoscale structures by adding or removing material from macroscale components
- molecular manufacturing to build systems from the atomic or molecular level (Nelson and Shipbaugh, 1995)

Biotechnology

Biological manufacturing processes take place under ambient conditions and generate very little waste. Biological systems are amazingly adept at exerting precise control over molecular synthesis and assembly processes to produce a wide range of components from a limited number of constituent materials. Some bioprocesses that could lead to revolutionary advances in manufacturing are listed below (NRC, 1994):

- biosynthetic pathways to genetically engineered protein polymers
- biological surfactant-based self-assembly processes that are effective in the 1 nm to 1,000 nm range
- methods of coupling synthesis and self-assembly processes to produce oriented and functionally-graded structures
- cell seeding and tissue engineering to enable *in vitro* production of skin and membranes
- biomineralization processes, including vesicle-mediated multicomponent processing

Process innovations in all aspects of manufacturing are likely to be incremental. However, breakthrough technologies in specific business sectors will also drive changes in ways that are difficult to predict. In nanotechnology and biotechnology, advances in basic sciences have provided the foundations for visions of "leapfrog" innovations. It is not a question of whether or not these technologies will have a significant effect on manufacturing, but rather when and how the effects will be felt.

3

Priority Technologies and Supporting Research

Manufacturing enterprises will require new capabilities to meet the grand challenges identified in Chapter 2. This chapter builds on the enabling technologies identified in Chapter 2 and describes the technologies that have the greatest potential to furnish these capabilities. Research opportunities to develop the priority technologies are also described.

Priority technologies and research opportunities were identified on the basis of the workshop, the Delphi survey, briefings by technology experts, and committee deliberations based on the following criteria:

- Was the technology identified as a high priority technology in the Delphi survey?
- Was the technology identified as a high priority technology at the workshop?
- Is this a primary technology for meeting one of the grand challenges?
- Does the technology have the potential to have a profound impact on manufacturing?
- Does the technology support more than one grand challenge?
- Does the technology represent a long-term opportunity (i.e., is the technology not readily attainable in the short term)?

The 10 technology areas selected as the most important for meeting the grand challenges are listed below (not in priority order):

- adaptable, integrated equipment, processes, and systems that can be readily reconfigured

- manufacturing processes that minimize waste production and energy consumption
- innovative processes to design and manufacture new materials and components
- biotechnology for manufacturing
- system synthesis, modeling, and simulation for all manufacturing operations
- technologies that can convert information into knowledge for effective decision making
- product and process design methods that address a broad range of product requirements
- enhanced human-machine interfaces
- educational and training methods that would enable the rapid assimilation of knowledge
- software for intelligent systems for collaboration

Examples of long-term research opportunities that will support the development of each technology to meet the grand challenges for 2020 are described in this chapter.

RECONFIGURABLE MANUFACTURING SYSTEMS

Adaptable, integrated equipment, processes, and systems that can be readily reconfigured for a wide range of customer requirements for products, features, and services is a priority technology. Hardware and software components, subprocesses, and subsystems will have to be adaptable and linked in easily programmable ways into higher-level processes and systems that span the entire product/service life cycle. Research opportunities to support the development of adaptable and reconfigurable manufacturing processes and systems fall into five broad areas: (1) processes and tooling, (2) theoretical foundations, (3) new manufacturing systems, (4) modeling and simulation, and (5) control and communications concepts.

Manufacturing Processes and Tooling

Adaptable, reconfigurable manufacturing processes and tooling include programmable, net-shape forming processes (e.g., free-form manufacturing concepts) that do not require hard tooling. Process technologies derived from rapid prototyping are particularly promising. Processing methods that can be readily reconfigured include nanofabrication concepts for manufacturing materials and components directly from molecular building blocks. Bioprocessing (i.e., combining molecular constituents to produce a range of products that vary in function and performance) could provide a model for manufacturing directly from molecular building blocks. Ultimately, a library would be developed of reusable processes and subprocesses for building and reconfiguring manufacturing systems (analogous to object-oriented programming in software development).

Tooling concepts range from modular tools, which would enable rapid changes in tooling within a process line, to a new tooling paradigm, in which hard tooling is replaced by software that defines the size, shape, and molecular constituents of a product.

Theoretical Foundations

The theoretical foundations for adaptable, reconfigurable manufacturing processes include the scientific basis for manufacturing processes on which models are based. Ultimately, simulations of manufacturing systems would be based on a unified taxonomy for process characteristics that include human characteristics in process models. Other areas for research include a general theory for adaptive systems that could be translated into manufacturing processes, systems, and the manufacturing enterprise; tools to optimize design choices to incorporate the most affordable manufacturing approaches; and systems research on the interaction between workers and manufacturing processes for the development of adaptive, flexible controls.

New Manufacturing Systems

New manufacturing systems will be required for adaptable and reconfigurable manufacturing processes that can meet the changing demands of the marketplace. Research in self-organizing manufacturing systems could include the development of autonomous manufacturing modules; bioprocessing technology; chaos theory; holonics[1]; and new concepts and models for partitioning manufacturing equipment, tools, human/organizational resources, and software systems. Finally, the development of new manufacturing systems must include a taxonomy and metrics for manufacturing systems that address open system architectures, mass customization, optimized system value, maximum use of available assets (including human, intellectual, and knowledge assets), and environmental impact.

Modeling and Simulation

Modeling and simulation capabilities for evaluating process and enterprise scenarios will be important in the development of reconfigurable enterprises. Simulations will have to be based on a systems view of the entire enterprise, including markets, workers, and cross-disciplinary interactions. Simulations will

---

[1] *Holonics* is a theory of organization and management that describes systems made up of interacting, self-similar units (*holons*). A similar, but less rigorous, concept is networked, autonomous, distributed units.

have to be designed to optimize human interaction, human and machine learning, and real-time information acquisition and analysis. Virtual prototyping of manufacturing processes and systems will enable manufacturers to evaluate a range of choices for optimizing their enterprises. Promising areas for the application of modeling and simulation technology for reconfigurable systems include neural networks for optimizing reconfiguration approaches and artificial intelligence for decision making.

Control and Communication Concepts

Processes that can be adapted or readily reconfigured will require flexible sensors and control algorithms that provide precision process control of a range of processes and environments. Reconfiguration of communications and control systems will rely on a common programming and control architecture, as well as "flexible" and "adaptive" software that does not require reprogramming but does provide operators with sufficient real-time information about the process to allow effective intervention, troubleshooting, and control.

WASTE-FREE PROCESSING

Manufacturing processes that minimize waste production and energy consumption is a key technology for the future. Manufacturing that does not damage the environment will be facilitated by manufacturing processes that do not create waste (e.g., free-form fabrication instead of material removal operations) or processes that create waste that can be used as feedstock in complementary manufacturing operations and, therefore, create no downstream waste. Processes that minimize energy consumption (e.g., processes with room-temperature bonding rather than high-temperature curing) will conserve resources and reduce costs and will also reduce indirect environmental effects from energy production.

Research in two principal areas will be required to meet the ultimate goal of waste-free processing—(1) waste reduction and utilization and (2) product design and analysis, including materials and process selection.

Waste Reduction and Utilization

The most effective way to reduce the environmental impact of manufacturing will be to use processes that do not produce by-products. Potential areas for research include net-shape processes (including net-shape forming, casting, and direct deposition), new routes in chemical synthesis that reduce or eliminate reaction by-products, and biological building processes. Incremental improvements in processes have been made, but the committee believes that breakthroughs in new process technologies will be necessary to approach waste-free processing goals.

Another key research area for reducing waste is processes that reuse by-products, either by recycling "home scrap" (use within the same factory) or by using process waste as a feedstock for another product line. A database will have to be developed for multiple material uses to match waste streams with potential users.

Product Design and Analysis

The production of sustainable products that have no detrimental effects on the environment throughout their life cycles will require advances in design tools and a philosophy based on concepts such as "design for reuse," which involves recovering major components or subsystems and reusing them instead of discarding them, and reprocessing materials and components, which involves re-manufacturing and upgrading products instead of discarding them. The development of modeling capabilities to minimize life cycle costs, including financial, resource, and environmental costs, will also be necessary.

NEW MATERIALS PROCESSES

Innovative processes to design and manufacture new materials and components will enable the manufacture of innovative, customized, waste-free products. The goal will be to develop new classes of materials with extraordinary physical properties (e.g., strength, wear characteristics, and electromagnetic properties). With miniaturization, new classes of "intelligent" products will be produced, but miniaturization at submicron scales will require materials with properties that can be controlled at the molecular scale. The processes for designing new materials and their components, especially submicron-sized components, will require design methodologies based on atomic and molecular physics and chemistry. In many cases, the materials will be organic, and the design methodologies will be biologically based. The processes for manufacturing these materials and components may require manipulation at the atomic level (nanofabrication), processes akin to gene splicing, and perhaps biological processes.

Research opportunities to support the development of processes to produce new classes of materials with extraordinary properties fall into three broad areas—innovative processing, design and analysis methods, and theoretical foundations.

Innovative Processing

Innovative processing methods include nanofabrication and improved net-shape processes. The committee believes that nanofabrication (nanoscale technology for fabrication) is an exciting leapfrog technology that could revolutionize manufacturing. The key technologies include nanomachining (e.g., nanolithography, abrasive ultraprecision finishing, and placement of atoms or molecules using techniques such as atomic-force microscopy and scanning tunneling microscopy), chemical-physical processing (e.g., molecular self-assembly, self-organizing structures, and ultrafine particle production), and bio-processing (described in the following section) (Nelson and Shipbaugh, 1995). Significant advances will have to be made in process measurement and control technologies, as well as in the fundamental understanding of processes to support design and modeling capabilities, before the promise of nanofabrication technology can be realized.

Programmable, net-shape forming processes will enable the development of adaptable, reconfigurable processing methods. Once products can be produced directly from a digital description without hard tooling, the development of cost-effective, customized, small-batch production processes with near-zero waste will become feasible.

Another research area is the development of measurement and control technologies (e.g., scanning tunneling microscopy, virtual reality, and feed-forward controls) that are applicable at submicron size scales. Ultimately, with the development of design, processing, and sensing and control technology with precise control of processes at all size scales—from the molecular level to the macro level—defect-free structures will be producible. The durability and reliability of these products could be far beyond those of current products.

Design and Analysis Methods

The development of innovative processing capabilities will require new concepts for life cycle material design. Potential advances include methods of designing and analyzing complex systems, such as "smart" materials (materials that can adapt to changing service requirements), biomimetic materials (materials based on biological models), and functionally gradient materials.

Theoretical Foundations

Manufacturing enterprises that apply advances in innovative materials processes will require a sound theoretical understanding of the processes and of materials performance. This will require capabilities for measuring and characterizing materials at extremely small size scales, design materials and components based on first-principles understanding, and precisely controlled processes and materials structures. Moreover, manufacturing enterprises will need technologies to collect, analyze, store, and use information based on performance and characterization experience to validate theoretical models.

BIOTECHNOLOGY FOR MANUFACTURING

Biotechnology for manufacturing has the potential to lead to revolutionary advances in innovative new products and manufacturing processes. Research would be based on an understanding of the precision and flexibility of biological processes and on finding ways to address their fundamental weaknesses (slow processes and the limited range of available materials).

New bioinspired and bioderived products will include biomemory and logic devices that can take advantage of the ways that biological organisms recognize environmental stimuli, learn, and adapt to changes; unique materials based on biological structures; and durable ultrasoft membrane materials.

Processing advances could include the fabrication of parts and assemblies with design enzymes, tissues, and biocatalysts; self-organizing manufacturing systems; and the genetic engineering of biological feedstocks to produce novel, tailored materials.

ENTERPRISE MODELING AND SIMULATION

Modeling and simulation for all operations of a manufacturing enterprise could enable the simulation of any operation, which could then be used for making decisions based on alternative scenarios. Detailed models of manufacturing enterprises—made up of integrated submodels describing the entire product/service life cycle—could be used for real-time control of all levels of manufacturing (from the manufacturing cell or factory floor to the globally distributed extended enterprise). Models and simulations should include descriptions of the interactions between people and between people and machines.

Research opportunities to support the development of modeling and simulation capabilities fall into two broad areas—(1) communications and information technology and (2) modeling tools.

Communications and Information Technology

Enterprise models will require the development of unified communication methods and protocols for the exchange of information, which will be used for the integration of process submodels of all levels of manufacturing enterprises, from individual human and process operations to distributed enterprises. Rapid communications will be required to support concurrent design and manufacturing. Unified methods and protocols should include a unified taxonomy, metrics for optimization, and identification of manufacturing primatives (basic operations of a manufacturing enterprise).

Significant advances in software will be needed for the integration of the whole range of submodels included in the enterprise model. Because software models will present an incomplete view of a dynamic enterprise, they will have to be adapted to incorporate new knowledge either through human intervention or machine intelligence. Key research areas include formalized representations of process knowledge to translate fundamental process information and design information for use in a variety of environments, reusable software modules, and enterprise models that incorporate new knowledge, and applications of artificial intelligence for flexible decision-making modules.

Developments in information technology will be required to support enterprise modeling. Promising research topics include planning tools for real-time decision making; representation of difficult abstractions and perspectives (e.g., value judgments); and display concepts that represent a large number of variables (e.g., information sources, content, reliability, robustness, degree of certainty, and application).

Modeling Tools

Research in enterprise modeling tools will include "soft" modeling (e.g., models that consider human behavior as an element of the system and models of information flow and communications), the optimization and integration of mixed models, the optimization of hardware systems, models of organizational structures and cross-organizational behavior, and models of complex or nonlinear systems and processes.

INFORMATION TECHNOLOGY

Converting information into knowledge for effective decision making is a priority technology. Integrated information technologies will be used to identify the information required for a specific decision, synthesize the information from distributed sources, filter out extraneous information, and present the information so that it can be used easily and immediately. The information system architecture should include semantics, protocols, and algorithms for conveying, filtering, and fusing data and information so that people can use the information for decision making.

Research opportunities to support the conversion of information to knowledge for decision making fall into three broad areas: (1) information synthesis, (2) presentation, and (3) architecture.

Information Synthesis

Future information systems will have to be able to collect and sift through vast amounts of information. Potential research areas for information synthesis include situation theory, human memory relational systems, and human-machine transformation technologies (e.g., from speech to text or from mind to computer). Promising research areas for filtering information include neural networks for interpreting data, case-based reasoning, artificial intelligence, intelligent agents for gathering knowledge ("knowbots"), and search engines based on "soft" semantics. Another research area is the development of methods to consider conflicting perspectives on causes and potential solutions to problems.

Information Presentation

Research and development will be needed on presentation methods for information systems that can present complex process variables and their relationships in forms that people with varying skills, capabilities, and backgrounds can use easily for decision making. Presentation technologies will have to allow for multiple levels of analysis, provide contextual information to facilitate accurate interpretation, and be customizable to individual preferences.

Information Architecture

Changes in the type and amount of information that manufacturing enterprises will use will require changes in the structure of knowledge databases. Research is needed to construct databases that include representations of cultural context, biotechnology architecture, the storage of knowledge (analyzed information rather than information), human behavior, manufacturing-oriented knowledge, and metastructures for "uncertainties" (e.g., degree of automation vs. human intervention).

PRODUCT AND PROCESS DESIGN METHODS

Product and process design methods that address a broad range of product requirements will be priority technologies for 2020. General purpose, modular design methods and tools could be used to meet a wide range of rapidly changing customer requirements. The methods and tools should accommodate scaleable and parametrically-defined families of products and processes; single, customized products; and mass-produced products. Design methods will also have to consider concepts and processes for a variety of materials, constructions, environmental conditions, and unique functional requirements, which might be thought of as "platforms" on which designs could be compiled from modular component and subsystem designs or edited from generic master designs.

The design system and tools should provide for complete simulations of products and enterprises and should integrate input from customers and workers, who will be integral members of the design team. Design tools should enable the enterprise to move directly from a digital product description to the development of production processes and tools.

Design methods will have to consider reconfigurations of products and processes, concurrent designs of products and production processes, optimized life-cycle costs, modular assembly, robust production processes, product flexibility, and social and environmental goals.

ENHANCED MACHINE-HUMAN INTERFACES

Enhanced human-machine interfaces between people, equipment, and information technology will be essential for manufacturing in 2020. Communications must be semantically correct, consider differences in human languages and cultures, and convey intention as well as facts. The interfaces must include all appropriate media for communication, building on the repertoire of technologies used today for virtual reality. Ideally, interfaces will be adaptive and customizable (i.e., they will be able to improve communications with specific individuals as they use the interfaces). Research opportunities fall into two principal areas: technical advances for the physical interface and learning technologies to enhance worker performance.

Seamless human input technologies could include a range of topics, from voice synthesis and control to full sensory input to direct mind-machine interfaces. Research on man-machine interfaces could include remote control for globally distributed enterprises, technologies that simplify and display large amounts of process data, interfaces that compensate for physical disabilities, and "smart" process algorithms.

Research on learning and design processes that will enhance worker performance include neural networks learning theories, decision support tools that are integrated with manufacturing operations and equipment, new techniques in education and cognitive science, training with simulations/virtual reality, and situation theory. In addition research will be needed to develop technologies for continuous learning by individuals and teams and collaborative design tools to allow people with different skills, education, cultural backgrounds, and organizational status to participate in the design process.

WORKFORCE EDUCATION AND TRAINING

Educational and training methods that would enable workers to assimilate knowledge to improve their effectiveness are priority technologies. Constant changes in manufacturing will place extreme demands on people to acquire and use new knowledge. Education and training technologies based on learning theory and the cognitive and linguistic sciences could provide knowledge in formats that could be used easily by a wide spectrum of individuals. These learning technologies will be supported by information technology for interactive, multimedia, distance learning, and information sciences for filtering and fusing knowledge for specific applications.

Long-term research will be based on changes in technologies that are available to educators (e.g., the transition to computer-based training) to teach people quickly in remote locations. The way people are educated and trained will change as enterprises become global, as jobs and skills change, and as new technology and processes are introduced.

Research opportunities include the development of tools that are not language or culturally dependent; technologies that can capitalize on advances in the cognitive sciences; interactive techniques, including simulation and virtual reality; and learning modules that can be adapted and tailored to meet individualized educational needs.

SOFTWARE FOR INTELLIGENT COLLABORATION SYSTEMS

The final priority technology is *software for intelligent systems for collaboration*. Intelligent systems for collaboration will enable people around the world, who have different functional expertise, communicate in different languages, and come from different cultures, to collaborate and interface through automated processes and machines. Collaboration systems will incorporate human-machine interfaces that can adapt to the user's expertise, language, and culture. They will also incorporate algorithms and methodologies for solving problems and facilitating organizational interactions.

The new tools will have to accommodate completely transparent remote interaction, including conferences, enterprise collaborations, and process controls. Long-term research goals include the development of protocols for group communication; network protocols specific to manufacturing (e.g., standards and protocols for the exchange of electronic data); methods and standards for controlling processes in a distributed enterprise; and methods for sharing enterprise and process knowledge.

Research on collaboration software should include human interaction interfaces based on models of human interaction dynamics that can represent human behavior and characteristics. The goal will be to provide a virtual space for collaboration that compensates for differences in skills, languages, cultures, organizational status, and terminology. The participation of educators and social and behavioral scientists will ease the transition to these kinds of interactions, which are likely to make many people uncomfortable.

MEETING THE GRAND CHALLENGES THROUGH TECHNOLOGY

Each priority technology is matched with prioritization factors in Table 3-1. All of the priority technologies would provide an enabling capability for meeting at least one of the grand challenges for manufacturing. The key technologies for each grand challenge are shown in Table 3-2. Descriptions of the enabling capabilities of each technology for each grand challenge are shown in Tables 3-3 through 3-8.

TABLE 3-1 Applicability of Evaluation Criteria to Priority Technology Areas[a]

Priority Technology	Criterion				
	Priority in Delphi Survey[b]	Priority in Workshop	Primary Technology for a Grand Challenge	Profound Impact on Manufacturing	Key Technology for Multiple Grand Challenges[c]
Adaptable and reconfigurable systems	1	✓	✓	✓	✓(5)
Waste-free processes	2	✓	✓		
New materials processes (e.g., submicron and nanoscale manufacturing)	7	✓	✓	✓	✓(2)
Biotechnology for manufacturing	8	✓	✓	✓	✓(2)
Enterprise modeling and simulation	2	✓	✓	✓	✓(6)
Information technology	6	✓	✓	✓	✓(6)
Improved design methodologies	11				✓(4)
Machine-human interfaces	5		✓	✓	✓(3)
Education and training	17	✓			✓(2)
Collaboration system software	21	✓			✓(2)

[a]All of the selected technologies represent long-term opportunities.
[b]Priority ranking based on voting by Delphi survey respondents (see Appendix B).
[c]Number of Grand Challenges where technology area was identified as applicable (see Table 3-2).

TABLE 3-2 Applicability of Priority Technology Areas to the Grand Challenges

	Grand Challenges					
Priority Technology	Concurrent Manufacturing	Integration of Human and Technical Resources	Conversion of Information to Knowledge	Environmental Compatibility	Reconfigurable Enterprises	Innovative Processes
Adaptable and reconfigurable systems	✓	✓	✓		✓	✓
Waste-free processes				✓		✓
New materials processes (e.g., submicron and nanoscale manufacturing)				✓		✓
Biotechnology for manufacturing				✓		✓
Enterprise modeling and simulation	✓	✓	✓	✓	✓	✓
Information technology	✓	✓	✓	✓	✓	✓
Improved design methodologies	✓			✓	✓	✓
Machine-human interfaces		✓	✓		✓	
Education and training		✓	✓			
Collaboration software systems	✓				✓	

TABLE 3-3 Enabling Capabilities for Concurrent Manufacturing (Grand Challenge 1)

Technology	Enabling Capabilities
Enterprise modeling and simulation	This is the primary technology for meeting this challenge. It provides the basis for understanding the interactions between the various entities of manufacturing enterprises. It enables the application of solutions that are optimized for the enterprise as a whole. It also enables the design of effective communication and information exchange systems.
Information technology	True concurrence requires more than the exchange of information. The knowledge exchanged must enable decisions based on the exchange. This technology will permit the recipient to convert diverse sources of information into knowledge that can be readily used. Without this technology, the integration of the diverse functions of the enterprise could result in gridlock rather than integration.
Improved design methodologies	This technology will allow close cooperation between product design and production and the simultaneous design of products and processes. It will also incorporate all of the parameters that describe the impact of the product design on all aspects of the enterprise (e.g., environmental, support, contractual, financial) into the initial design process.
Collaboration software systems	A 2020 concurrent organization will not only require the ability to exchange information and knowledge, but will also require effective interaction between entities. This technology will facilitate interaction by overcoming differences in context, culture, terminology, and language.

TABLE 3-4 Enabling Capabilities for Integrated Human and Technical Resources (Grand Challenge 2)

Technology	Enabling Capabilities
Machine-human interfaces	Advanced machine interfaces will enable people to make independent decisions that will enable them to control production processes. People will be able to understand the ramifications of process changes on products and on the manufacturing system. Enhanced interfaces will facilitate conceptualization and provide information in a context that promotes understanding.
Adaptable and reconfigurable systems	The development of processes that can be reconfigured easily will empower people to make changes on the floor to meet changing demands. The team of 2020 will require new levels of interaction with technology. Trying out new processes and prototyping new methods will not be fast enough. With virtual reality software, people will be able to determine what will work and what won't and quickly understand the impact of process changes. Highly adaptable teams trained in the use of a wide variety of tools will implement changes in the enterprise. System tools will be used routinely by team members to measure projected improvements and the consequences of reconfigured processes.
Enterprise modeling and simulation	This technology will link the production worker, the production process, and the rest of the enterprise. It can provide feedback on the negative, as well as positive, effects of a worker's actions. Workers will be able to optimize processes and make decisions based on enterprise considerations.
Information technology	This technology will systematically and consistently present knowledge so that it facilitates work. Information technologies will automatically discard irrelevant information.
Education and training	This technology will enable people to acquire and use knowledge quickly and effectively, making workers more confident and better able to respond to new circumstances.

TABLE 3-5 Enabling Capabilities for Converting Information to Knowledge (Grand Challenge 3)

Technology	Enabling Capabilities
Information technology	This is a primary technology to meet this challenge. Technology and the competitive environment will continue to change very quickly. The underlying infrastructure will require system architectures and algorithms and methodologies for acquiring information and converting it into immediately usable knowledge.
Enterprise modeling and simulation	This technology will guide the processes for converting information into knowledge and for conceptualizing manufacturing functions and operations. The technology is essential to combining information from many sources into a consistent description of the enterprise and its operations. Effective decisions will depend on predictions, perhaps even optimizations, of system behaviors.
Machine-human interfaces	This technology will enable individuals to access the information and knowledge within the enterprise's systems. Individuals will participate in synthesizing knowledge for application in manufacturing operations.
Education and training	This technology will enable workers to participate in the transformation of information into useful knowledge. As the ultimate decision makers, workers will be required to acquire, accept, and process information and knowledge in ways that can be used in manufacturing operations.

TABLE 3-6 Enabling Capabilities for Environmental Compatibility (Grand Challenge 4)

Technology	Enabling Capabilities
Waste-free processes	This is the primary technology to meet this challenge. The objective of the new or modified processes must always be to produce no waste of any kind, to consume the minimum amount of energy, and to do both economically.
New materials processes (e.g., submicron and nanoscale manufacturing)	One way to reduce waste is to use free-form fabrication with tolerances that do not require material removal. Another is to build products using materials with environmentally favorable physical characteristics, such as very lightweight, but very strong, structural components.
Biotechnology for manufacturing	Biotechnology offers the possibility of using renewable biological processes for manufacturing, to manufacture biologically-defined products, and to create only nonpolluting, biodegradable wastes.
Improved design methodologies	With this technology, environmental considerations, energy utilization, and waste minimization can be considered early in product and process design. Design methodologies that include these factors as part of the trade-off criteria will enable the design of affordable products that have minimal adverse environmental impacts.

TABLE 3-7 Enabling Capabilities for Reconfigurable Enterprises (Grand Challenge 5)

Technology	Enabling Capabilities
Adaptable and reconfigurable systems	This is the primary technology to meet this challenge. The rapid reconfiguration of enterprises will require that the underlying equipment, manufacturing and business processes, and manufacturing systems all be rapidly reconfigurable. Equipment and unit processes must also be easily integrated into macroprocesses and systems.
Enterprise modeling and simulation	Rapid, virtual prototyping based on advanced modeling and simulation of complex manufacturing processes and systems will enable "just-in-time" reconfiguration decisions for products and physical processes; appropriate business processes; and enterprise design, organization, operations, and control.
Information technology	This technology will provide the underlying infrastructure, architecture, algorithms, and methodologies for acquiring information and converting it into immediately usable knowledge. In an environment where product and process technology are changing very quickly, knowledge that can be used in real time for operations and decision making will be crucial. Information will be synthesized from diverse sources.
Improved design methodologies	This technology will provide design methodologies that can be quickly adapted to accommodate significant changes in requirements. The capability to reconfigure designs quickly will be the basis for reconfigurations throughout the enterprise. In most competitive situations, there will be little time for constructing new design methodologies *ab initio*.
Machine-human interfaces	This technology will enable timely decisions on complex manufacturing issues at any level of the enterprise. People will participate in key decisions about design and operations when equipment, processes, systems, or the enterprise itself are reconfigured. People will make key decisions based on knowledge provided by automated machines and systems.

continued

TABLE 3-7 *Continued*

Technology	Enabling Capabilities
Education and training	New knowledge delivery systems will facilitate rapid learning for manufacturing applications. Individuals involved in the planning and operations of reconfigurations will be asked to make quick, accurate decisions. Their ability to do so will depend on their ability to learn.
Collaboration software systems	Reconfiguration at any level will involve many people interacting with each other and with machines. This technology will be able to accommodate teams whose members are separated geographically and have widely different functional backgrounds, skill levels, languages, and cultures.

TABLE 3-8 Enabling Capabilities for Innovative Processes (Grand Challenge 6)

Technology	Enabling Capabilities
New materials processes (e.g., submicron and nanoscale manufacturing)	With this technology, new materials with unusual properties (e.g., room-temperature superconductivity, electromagnetic properties confined to submicron domains, and unidirectional heat flows) will make possible new classes of products or radical re-engineering of traditional products. Technology to create these new materials and then manufacture them in bulk will be required to realize this potential. Manufacturing components using these new materials will also require new processes.
Biotechnology for manufacturing	This technology will enable a special class of manufacturing: biological processes to manufacture new raw materials and finished components with biologically defined properties and shapes. The technology will enable new products and products using hybrid materials.
Adaptable and reconfigurable systems	This technology will make possible programmable equipment, processes, and systems that can be used to create a broad range of products rapidly and with minimal changeover costs. Affordable, one-of-a-kind products will be quickly produced to meet specific customer requirements.
Improved design methodologies	This technology will provide new design methodologies that can be quickly adapted to accommodate major changes in requirements. Timely, affordable, one-of-a-kind products will require rapid product and process designs.

4

Preparing for 2020

Manufacturing in 2020 will be exciting, dynamic, and competitive. With the emergence of billions of new consumers into the "developed" world, the emphasis on education, the pressure to raise or maintain living standards while consuming fewer resources, and the global availability of knowledge, manufacturers will have unprecedented market opportunities but will also be subject to unprecedented competitive pressures. Chapter 2 identifies six grand challenges that manufacturers will have to meet to thrive under these conditions and outlines technical opportunities for meeting them. Chapter 3 describes 10 priority technology areas for addressing the grand challenges and outlines research opportunities related to the priority technology areas. The committee recommends that long-term manufacturing research focus on developing capabilities in the priority technology areas to meet the grand challenges.

This chapter highlights the committee's general findings:

- Many of the areas for research are crosscutting areas, that is, they are applicable to several of the priority technologies identified in Chapter 3.
- Two important breakthrough technologies—submicron manufacturing and simulation and modeling—will accelerate progress in addressing the grand challenges.
- Substantial research is already under way outside of the manufacturing sector that could be focused on manufacturing applications.
- Progress toward the goals recommended in the Next Generation Manufacturing study (NGM, 1997) on the needs of the next decade will also contribute to meeting the longer-term grand challenges for 2020.

- Because manufacturing is inherently multidisciplinary and involves a complicated mix of people, systems, processes, and equipment, the most effective research will also be multidisciplinary and grounded in knowledge of manufacturing strategies, planning, and operations.

The committee's findings and recommendations are described in more detail below.

CROSSCUTTING RESEARCH

Table 3-2, which relates the priority technologies to the grand challenges, shows that the development of the priority technologies will affect several of the grand challenges. Many of the research areas described briefly in Chapter 3 can potentially contribute to the development of more than one priority technology. This has both advantages, in that research resources can be used more efficiently, and disadvantages, in that results may not necessarily apply to all of the priority technologies.

The following examples illustrate how research could be applicable to more than one technology. First, the development of adaptive, reconfigurable equipment, processes, and systems will enable the rapid reconfiguration of enterprises to meet competitive pressures but will also improve the integration of human and technology resources, enterprise-wide concurrency, and the development of revolutionary processes. Second, research on modeling and simulation will help meet the challenges for enterprise-wide concurrency, the utilization of human and technological resources, the conversion of information to knowledge, and the rapid reconfiguration of manufacturing enterprises. Finally, research on information technology will help to meet all of the grand challenges. Information technology is the primary technology for converting information to knowledge and will be a key technology for concurrency, the integration of human and technical resources, and the rapid reconfiguration of enterprises.

Recommendation. Establish priorities for long-term research with an emphasis on crosscutting technologies, i.e., technologies that address more than one grand challenge. Adaptable and reconfigurable manufacturing systems, information and communication technologies, and modeling and simulation are three research areas that address several grand challenges.

BREAKTHROUGH RESEARCH

The committee believes that technological breakthroughs in two areas—innovative submicron manufacturing processes and enterprise modeling and simulation—would have a profound impact on manufacturing of the future.

Submicron Manufacturing

Submicron manufacturing promises to provide economic solutions to meeting increasingly demanding customer needs and, at the same time, decreasing time to market, energy consumption, and environmental costs. Manufacturing at the submicron level has four important aspects—evolutionary advances in (1) miniaturization and (2) microelectromechanical systems (MEMS), as well as revolutionary advances in (3) nanofabrication and (4) biotechnology.

There has been a steady trend toward miniaturizing manufactured components. A good example is the progression from vacuum tubes and discrete transistors to the very dense integrated circuits manufactured today. Integrated circuits contain structures, produced in layers using photolithograpic processes, with features on the order of a micron or less in size. For compelling economic reasons, the semiconductor industry continues to reduce the dimensions of integrated circuits. The proliferation of more and more powerful, but smaller and smaller, intelligent systems will lead to advances that will be crucial for meeting several of the grand challenges for manufacturing in 2020.

MEMS use sensors, actuators, and other electromechanical structures with dimensions on the order of microns (NRC, 1997). Like integrated circuits, MEMS are produced using the batch-processing capabilities of semiconductor processing. In fact, MEMS can be part of integrated circuits that combine machine intelligence with electromechanical action.

The ultimate in submicron manufacturing is nanofabrication, specifically molecular nanotechnology (MNT), in which individual atoms and molecules are manipulated to form materials and structures. The consensus among MNT researchers is that, in principle, a wide range of molecular structures can be produced cost effectively. MNT could enable the production of new materials with specific properties tailored for given applications, properties that could be varied as structures are built up to produce functionally gradient materials. In addition, materials and structures with dramatically improved properties could be produced with no waste. Costs for self-replicating materials manufactured by MNT could be reduced to competitive levels by 2020. If costs are competitive, MNT will have far-reaching implications for waste-free manufacturing of very light weight, strong microstructures and macrostructures.

One important form of self-replication at the molecular level that occurs naturally is controlled by DNA and cellular processes. Biotechnology has already progressed to the point that genes can be manipulated. By 2020, a substantial technology will have been developed for the production of biological materials, the replication of biological materials, and the formation of structures from biological materials. The interrelationship between bioprocessing and MNT could lead to the production of hybrid structures that combine DNA and machine intelligence with biological and nonbiological materials.

Modeling and Simulation of Manufacturing Systems

Meeting the grand challenges of concurrency in all operations (grand challenge 1) and rapid reconfiguration of manufacturing enterprises (grand challenge 5)—which include enterprise strategy, planning, and operations at one extreme and manufacturing cell operations at the other—will depend on accurate predictions and timely decisions based on modeling and simulation to develop virtual prototypes. Manufacturing systems in 2020 will be complicated, dynamic amalgams of human and machine intelligence, knowledge, materials, equipment, and processes. Operational decisions made at relatively low levels in the enterprise may have enterprise-wide consequences.

Two crucial elements are necessary for successful manufacturing systems models and simulations—a comprehensive set of models and human-machine interfaces that enable individuals to interact with the models for learning, planning, and manufacturing control. The semantics of manufacturing that encompasses all enterprise operations and functions within a globally distributed real (or virtual) manufacturing enterprise must be consistent across all levels, operations, and functions of the enterprise. Ideally, the semantics would support global multi-objective optimization of the enterprise and its operations; that is, it would be robust enough to be the basis for a *theory of manufacturing* and adaptable enough to support change.

Individuals will be critical components of any manufacturing system. Models and simulations must account for individuals from two points of view. First, the behavior and actions of individuals, as part of a manufacturing system, must be included in the models. This implies an understanding of how individuals relate to each other within the system, as well as an understanding of how individuals relate to equipment and processes (which may or may not be automated). Second, models and simulations must be described and delivered in a usable form to facilitate the decision or action that must be taken.

Including human behavior, with all of its vagaries of learning and communication styles and overtones of culture and language, will make modeling and simulation difficult. However, unless the human factor is included, the representation will be unrealistic.

Recommendation. Establish basic research focused on breakthrough technologies, including innovative submicron manufacturing processes and enterprise modeling and simulation. Focus basic research on the development of a scientific base for production processes and systems that support new generations of innovative products.

TAKING ADVANTAGE OF "DRIVER" TECHNOLOGIES

Some of the technology areas for meeting the grand challenges are being developed for other purposes. For example, information is a core technology that is applicable to grand challenges for concurrency in all operations (grand challenge 1), integration of human and technical resources (grand challenge 2), transformation of information into knowledge (grand challenge 3), and rapid reconfiguration of manufacturing enterprises (grand challenge 5). A very significant investment in the development of information technology is already being made to meet the needs of other sectors of the economy and will eventually lead to global systems that are interoperable at the level of communications systems and operating systems and that will enable advanced human-machine interfaces with auditory, visual, and tactile capabilities. However, information technologies that enable seamless, collaborative systems may not be useful for manufacturing without a further determination of how people, machines, and information technology can work together beneficially in manufacturing systems. Individuals in a specific linguistic and cultural situation must be able to communicate using the medium of information technology with machines, complicated manufacturing systems, and people in different linguistic and cultural situations.

Recommendation. Monitor research and development on technologies with significant investment from outside the manufacturing sector and undertake research and development, as necessary, to adapt them for manufacturing applications. Some applicable technologies are listed below:

- information technology that can be adapted and incorporated into collaboration systems and models through manufacturing-specific research and development focused on improving methods for people to make decisions, individually and as part of a group
- core technologies, including materials science, energy conservation, and environmental protection technologies

BUILDING ON NEXT-GENERATION MANUFACTURING

The Next Generation Manufacturing Project was a national, industry-led project conducted in 1995–1996. Nearly 500 people, mostly managers and technical experts from manufacturing companies, participated (NGM, 1997). The objectives of the project were (1) to develop a broadly accepted model of future manufacturing enterprises ("future" was defined as the next decade) and (2) to recommend actions that manufacturers, working individually and in partnership with government, industry, and the academic community, could take to attain "world-class" status.

The Next Generation Manufacturing Project defined a typical manufacturing company of the next decade and developed a framework for actions that would make U.S. companies globally competitive between now and 2010. Executives of leading companies first defined *pragmatic dilemmas* they face. Starting from this pragmatic base, they described key competitive *drivers*, identified the attributes of a successful company, and characterized the capabilities, or *imperatives*, required for companies to thrive. The project also recommended steps companies could take to achieve these capabilities.

One important recommendation was that manufacturers develop technology road maps to identify research and development that would support the transition of present-day companies to next-generation companies. A project led by Oak Ridge National Laboratory, called the Integrated Manufacturing Technology Roadmap Initiative, was established to address this recommendation in terms of information systems, modeling and simulation, manufacturing processes and equipment, and enterprise integration.

Most of the recommendations involved the development and implementation of new business practices and organizations or the application of existing technologies to advanced manufacturing. However, a few recommendations involved research and development. These recommendations are described below along with the committee's assessment of their applicability to manufacturing in 2020.

Develop Next-Generation Models and Assessment Capability

This recommendation focused on adapting existing models to develop an integrated reference set of multilevel models. These models would be used to facilitate the participation of companies in extended enterprises, to facilitate the transition of present-day companies into next-generation companies, and to educate company personnel. A complementary recommendation focused on tools for assessing a company's capabilities.

Operations Modeling and Simulation workshops were held at the Oak Ridge National Laboratory to follow up on this recommendation. The committee expects that the evolutionary advances in this area will be a valuable subset of the models and simulations that will be required in 2020 to support enterprise-wide concurrency. But revolutionary advances in communication standards and protocols, human-machine interfaces, and models and simulations that include human and organizational behavior will also be necessary for manufacturing to realize the 2020 vision of enterprise modeling and simulation.

Develop Systematic Processes for Capturing Knowledge and for Knowledge-Based Manufacturing

The goal of this NGM recommendation was the development of a usable repository of manufacturing knowledge that could be an easily accessible core for a knowledge base. The processes for capturing knowledge would conform to a consistent set of rules applicable across the entire product life cycle. People applying the knowledge would also be guided by consistent rules, possibly incorporated into automated systems.

The research necessary for fulfilling this recommendation could result in knowledge acquisition and delivery systems that could become the foundation for the committee's recommended research on converting information into knowledge and developing knowledge systems for rapidly reconfigurable processes and equipment in 2020.

Enable and Promote the Use of Modeling and Simulation

The goal of this recommendation was to advance the state of the art by establishing standards for the verification, validation, and accreditation of modeling tools and models (including geometric models, behavioral models, process models, and cost and performance models).

The direction for next-generation manufacturing was consistent with the goals for models and simulations in 2020. Fulfillment of this recommendation would provide fundamental building blocks for the dynamic models and "real-time" simulations of 2020. But, as described above, additional advances in communications, human-machine interfaces, and consideration of human and organizational behavior will be necessary to realize the 2020 vision of enterprise modeling and simulation.

Develop Intelligent Processes and Flexible Manufacturing Systems

The goal of this NGM recommendation was the development and establishment of a methodology for introducing intelligent processing into manufacturing systems. Intelligent processing would reduce the programming burden when product requirements, processes, and factory configurations must be changed. Intelligent processing systems would be able to adapt automatically or semi-automatically. Fulfillment of this recommendation would provide building blocks for the rapidly reconfigurable manufacturing enterprises of 2020.

INTERDISCIPLINARY RESEARCH

The committee believes that research related to manufacturing enterprises is inherently interdisciplinary and that the development of the priority technology areas for 2020 manufacturing will require an unprecedented commitment to multidisciplinary and collaborative research. The grand challenges, which reflect real-world complexities, are not amenable to single-discipline solutions. The working relationships between the physical science and engineering disciplines that have emerged in recent decades will have to be expanded to include mathematics, economics, enterprise management, computer science, philosophy, biology, psychology, cognitive science, and anthropology.

The manufacturing industry will have to (1) identify current real problems and forecast the problems enterprises will face in the future and (2) articulate these problems in terms that are accessible by academic and research organizations. At the same time, the academic and research community will have to (1) facilitate the formation of integrated teams and (2) articulate the technical results of research in terms that are accessible by industry leaders.

Recommendation. Establish an interdisciplinary research and development program that emphasizes multi-investigator consortia both within institutions and across institutional boundaries. Establish links between research communities in the important disciplines required to address the grand challenges, including all branches of engineering, mathematics, physics, chemistry, economics, management science, computer science, philosophy, biology, psychology, cognitive science, and anthropology.

Recommendation. Industry and government should focus interdisciplinary research and development on the priority technology areas. Some key considerations for the long-term are listed below:

- understanding the effect of human psychology and social sciences on decision-making processes in the design, planning, and operation of manufacturing processes
- managing and using information to make intelligent decisions among a vast array of alternatives
- adapting and reconfiguring manufacturing processes rapidly for the production of diverse, customized products
- adapting and reconfiguring manufacturing enterprises to enable the formation of complex alliances with other organizations
- developing concurrent engineering tools that facilitate cross-disciplinary and enterprise-wide involvement in the conceptualization, design, and production of products and services to reduce time-to-market and improve quality
- developing educational and training technologies based on learning theory and the cognitive and linguistic sciences to enhance interactive distance learning
- optimizing the use of human intelligence to complement the application and implementation of new technology
- understanding the effects of new technologies on the manufacturing workforce, the work environment, and the surrounding community
- developing business and engineering tools that are transparent to differences in skills, education, status, language, and culture to bridge international and organizational boundaries

MEASURING PROGRESS

One of the key factors in meeting the grand challenges will be monitoring the progress of technology development. The committee believes a detailed research agenda and timetable based on the grand challenges and priority technology areas for manufacturing in 2020 should be developed. However, detailed research agendas or timetables were beyond the scope of this study. Research road maps that could be used to monitor progress toward realization of the vision of manufacturing in 2020 should be established in follow-up technology seminars with focus groups exploring the priority technologies and potential research areas. Rather than trying to anticipate the advancements for a twenty-year period, the committee recommends that general long-term goals be established in each technology area and that detailed road maps be established for five-year "windows of commitment." This approach, similar to the approach of the Defense Advanced Research Projects Agency, would provide a reasonable time frame for technology incubation, with yearly reviews to monitor progress. At the end of the five-year period, goals and programs would be re-examined for the next five-year period. This approach would allow research efforts to be adapted to revolutionary advances and for unfruitful research directions to be reconsidered.

References

ACS (American Chemical Society). 1996. Technology Vision 2020: The U.S. Chemical Industry. Washington, D.C.: The American Chemical Society.

Adelson, M., and S. Aroni. 1975. Differential images of the future. Pp. 433–462 in The Delphi Method: Techniques and Applications, H. Linstone and M. Turoff, eds. Reading, Mass.: Addison-Wesley Publishing Company.

CAD/CAM Update. 1997. IBM and Daussault awarded Boeing CATIA contract. CAD/CAM Update 9: 1–8.

Computing Canada. 1997. Virtual development in practice. Computing Canada 23(26): 36.

Kaplan, R.S., and D.P. Norton. 1996. The Balanced Scorecard: Translating Strategy into Action. Cambridge, Mass.: Harvard Business School Press.

Nelson, M., and C. Shipbaugh. 1995. The Potential for Nanotechnology for Molecular Manufacturing. MR-615-RC. Santa Monica, Calif.: Rand Corporation.

NGM (Next-Generation Manufacturing). 1997. Next-Generation Manufacturing: A Framework for Action. Bethlehem, Pa.: Agility Forum.

NRC (National Research Council). 1998. Manufacturing Process Controls for the Industries of the Future. NMAB-487-2. Washington, D.C.: National Academy Press.

NRC. 1997. Microelectromechanical Systems: Advanced Materials and Fabrication Methods. NMAB-483. Washington, D.C.: National Academy Press.

NRC. 1996. Linking Science and Technology to Society's Environmental Goals. Washington, D.C.: National Academy Press.

NRC. 1994. Hierarchical Structures in Biology as a Guide for New Materials Technology. NMAB-464. Washington, D.C.: National Academy Press.

NRC. 1992. Beam Technologies for Integrated Processing. NMAB-461. Washington, D.C.: National Academy Press.

Reuters. 1997. Chrysler shaves 8 months off LH car development. Reuters Business Report August 4.

Sheng, P., and B. Allenby. 1997. Environmental Considerations in Manufacturing. Presentation at the Workshop on Visionary Manufacturing Challenges, National Research Council, Beckman Center, Irvine, California, April 1–3, 1997.

APPENDICES

APPENDIX

A

Summary of Workshop on Visionary Manufacturing Challenges

Irvine, California, April 1–2, 1997

WELCOME AND PURPOSE OF WORKSHOP

John Bollinger, Chair

John Bollinger welcomed the participants to the Workshop on Visionary Manufacturing Challenges and explained that for the next two days the participants would attempt to develop a vision for a small but critical aspect of the future. He noted that he could not think of a better day for the workshop to begin than April 1st. Bollinger expressed confidence that this vision would be pertinent to many changes in society between now and the years beyond 2020.

Bollinger defined the objective of the National Research Council Committee on Visionary Manufacturing Challenges, which had organized the workshop, as the identification of technologies and systems that are likely to be important for manufacturing in the decades after 2020 as a guide for funding current and future research. He said that the study would be based on the following premises:

- The manufacturing environment will continue to change rapidly.
- Competition will be intense.
- Dramatically new products and processes will emerge.
- New management and labor practices will emerge.
- Manufacturing will remain one of the principal means of creating wealth.

Bollinger told workshop participants that the study, which would be international in scope, would be informed by three sources: past studies, a Delphi-type survey, and this workshop. The challenge facing the workshop participants would be to identify "leapfrog" opportunities, to define the challenges for future manufacturing enterprises, and to define enabling technologies for meeting those challenges.

Bollinger quoted from a recent article by Peter Drucker in *Forbes* magazine describing his vision for 2050, in which he made the following predictions:

- The poor will rise up against the rich.
- Chinese clans will control world markets.
- Industry will be too dependent on computers.
- Academic institutions will be redundant.

Bollinger pointed out that approximately 50 percent of Drucker's predictions have been correct in the past and that greed is already rampant, China is the acknowledged new market horizon, and industry is becoming increasingly dependent on computers. Never before, however, has industry so emphatically asserted the necessity for employee training and education.

Bollinger emphasized that the ideas brought forward at the workshop need not be verifiable because the workshop was a vehicle for exploring the possibilities of the year 2020 and beyond, and participants were not necessarily expected to be right. He pointed out that a recent project, Next Generation Manufacturing (NGM), had focused on evolutionary transitions, ideas that could be conceived today and applied tomorrow based on existing initiatives. The purpose of this workshop, however, was to focus on the next century, to imagine the challenges and needs that could shape investment strategies for manufacturing research.

Finally, Bollinger described the workshop itself, which was divided into four sessions, each of which would begin with thought-provoking presentations. After the presentations, participants would be divided into small brainstorming groups, with committee members acting as facilitators. Each group was asked to select one person to act as a "reporter" and present the results of the discussions at the plenary session at the end of each day. Bollinger closed with the hope that participants would enjoy the workshop and thanked them for their participation.

WORKSHOP ORGANIZATION

Workshop participants (see Box A-1) were divided into six discussion groups with the goal of generating original ideas and new insights. The discussion groups were asked to consider the opening presentations as food for thought rather than as boundaries for their discussion. The groups met twice each day and presented the results of their discussions during the plenary sessions that followed. They were given specific questions to answer at each session. A committee member served as facilitator for each group.

After each group had restated the question and the objectives of the session, a brainstorming period ensued during which everyone provided ideas and suggestions without discussion. This material was then organized and prioritized for presentation by the reporter at the plenary session.

BOX A-1
Workshop Participants

Richard Altman, *Communication Design*
Debra M. Amidon, *Entovation International*
John Bollinger, *University of Wisconsin-Madison*
Steven J. Bomba, *Johnson Controls*
Philip Burgess, *Center for the New West*
Charles Carter, Jr., *The Association for Manufacturing Technology*
Nathan Cloud, *DuPont*
Thomas Crumm, *General Motors Corporation*
John Decaire, *National Center for Manufacturing Sciences*
Rick Dove, *Paradigm Shift International*
Gordon Forward, *Chaparral Steel*
Barbara Fossum, *University of Texas*
Donald Frey, *Northwestern University*
H.T. Goranson, *Sirius Beta*
David Hagen, *Michigan Center for High Technology*
William Hanson, *Massachusetts Institute of Technology*
David Hardt, *Massachusetts Institute of Technology*
George Hazelrigg, *National Science Foundation*
Robert Hocken, *University of North Carolina-Charlotte*
Richard Jarman, *Eastman Kodak Company*
Bill Kay, *Hewlett-Packard Company*
Richard Kegg, *Cincinnati Milicron, Inc.*
Louis Kiefer, *International Association of Machinists and Auto Workers*
Howard Kuhn, *Concurrent Technologies Corporation*
Eric Larson, *Rand Corporation*
Edward Leamer, *University of California at Los Angeles*
Ann Majchrzak, *University of Southern California*
Mike McEvoy, *Baxter International, Inc.*
Rakesh Mahajan, *DENEB Robotics, Inc.*
M. Eugene Merchant, *Institute of Advanced Manufacturing Sciences*
David Miska, *United Technologies Corporation*
Richard Morley, *Morley and Associates*
Richard Neal, *Lockheed Martin*
Woody Noxon, *CAM-I*
Leo Plonsky, *U.S. Navy Industrial Resources Support*
Lawrence Rhoades, *Extrude Hone Corporation*
Heinz Schmitt, *Sandia National Laboratories*
F. Stan Settles, *University of Southern California*
Paul Sheng, *University of California at Berkeley*
Wilfried Sihn, *Fraunhofer Institute for Manufacturing Engineering and Automation*
James Solberg, *Purdue University*
Brian Turner, *Work and Technology Institute*
Mauro Walker, *Motorola*
Kathryn Whiting, *Boeing Defense and Space Group*
Patricia Whitman, *Los Angeles County Office of Education*
Eugene Wong, *University of California at Berkeley*

PART I

Global Issues and Competition in 2020

DRIVERS FOR INDUSTRY IN 2020

Philip Burgess

Center for the New West, Denver, Colorado

Philip Burgess began by stating that forecasting is a tricky business and that the records show we're not very good at it. For example, Alexander Graham Bell predicted in 1887 that the telephone was such an important invention that "someday every community would have one." In 1889, Western Union decided not to purchase all of Bell's patents for $100,000 because they did not believe there was a market for this "electronic toy." In 1899, the U.S. Patent Office director, Charles Duell, stated that everything that could be invented had been invented. Wilbur Wright predicted in 1901 that humans would not fly for another 50 years. In 1903, Horace Rackham predicted that the horse was here to stay and that automobiles were just a fad, although he also bought stock in Ford Motor Company. In 1911, Ferdinand Foch said that, in his opinion, although "aeroplanes" were interesting they were of no military value. In 1927, Warner Brothers wondered who would want to hear actors talk. In 1943, Thomas Watson forecast a world market for about five computers. In 1977, Kenneth Olsen, founder and president of Digital Equipment Corporation, said no one needed to have a personal computer at home. In 1981, Bill Gates said that 640K would be enough memory for anyone. In 1989, Irving Fisher said that stocks had reached a permanently high plateau.

Burgess went on to say that major changes are occurring in the United States and worldwide and that he believes we are entering a new age, characterized by the growing importance of intellectual capital and its impact on all areas of life. He also believes we are entering a new economy, characterized by expanded global competition, with the focus on new methods of distribution and delivery and the integration of these functions with the manufacturing process. The social and political manifestations of this new regime include dramatic demographic shifts, democratization, decentralization, and other developments that will limit institutional power. For example, new technologies like the Internet will continue to empower people, thereby threatening institutional power.

Burgess believes that some of these social manifestations constitute a "value revolution," although he thinks "value restoration" might be a more descriptive term. He suggested that a new Luddite movement might be in the making. As evidence, he cited the recent controversy over cloning and noted that only one of the three major news magazines had focused on the promising aspects of cloning technology; the other two had focused on human cloning and other sensational aspects of the topic. He also cited a renewed interest in fundamental values around the world.

According to Burgess, the new regime will reward people and organizations that are fast, flexible, focused, customized, networked, and global. The broad forces at work are distributive, moving power and control from the center to the periphery. He believes that the United States is especially well suited to prosper in this new regime, which will include on-site manufacturing and the capability of producing customized products quickly. In contrast to the United States, the European Union will have problems in the new regime because it is a "mainframe" concept in a "PC" world and has created a new layer of centralized bureaucracy. None of the world's leading industries is headquartered in Europe.

Burgess calls the driving forces for change "TIDES of the Millennium": **T**echnology, **I**nternational commerce, **D**emography, **E**ntrepreneurship, and **S**tandards of living.

Technology. The importance of technology, which has been and will continue to be a driver, cannot be overemphasized. The technology-driven industries of the next century will be civil aviation, biotechnology, materials, microelectronics, computers and software, telecommunications, robotics, and machine tools.

International commerce. The Anglo-American way of doing business is being adopted worldwide, including accounting practices, advertising, corporate finance, business education, and business ethics. English is the language of commerce and diplomacy, and more Chinese are learning English today than there are Americans. The Anglo-American diaspora is larger than the Jewish diaspora and more influential than the Chinese diaspora of 55 million.

Demography. People are an economy's most important asset because only people have the ability to sense, judge, create, and build relationships. The United States has a big advantage because it is a magnet for immigration. First-generation immigrants from Taiwan, Yugoslavia, and Pakistan currently run six of the top fifteen corporations in southern California; three more are being run by second-generation immigrants. In Silicon Valley, one-third of the engineers is Asian. The United States has a huge asset in these people.

Entrepreneurship. The United States has one of the strongest family-based entrepreneurial cultures in the world, matched only by the Chinese, including the Chinese in Taiwan, Singapore, and Hong Kong. Today, the United States has 22 million business enterprises. Of these, only 14,000 have more than 500 employees. The action is therefore with small enterprises, which have accounted for 100 percent of net new job growth in the past seven years. In the economy of the future, many small and mid-sized enterprises (SMEs) will operate through networks, such as learning networks, intelligence networks, resource networks, distribution networks, co-marketing networks, co-production networks, and joint-procurement networks. Attempts will be made to drive cost out of the system. As an example of a seamless link between manufacturing, distribution, and delivery, Domino's pizza could be produced in mobile units, thereby reducing costs and speeding up delivery.

Standards of living. Standards of living are rising all over the world. The net result of this is positive. In the future, travel will increase, and because of higher per capita income, the investment in a clean environment will also increase. Tremendous new markets will open up for environmental technologies, new infrastructures will be built, and manufacturing will become even more important than it is today.

At this point in his talk, Burgess turned to a more in-depth discussion of technology, the first of his five TIDES. He cited a recent MIT study that identified the following major technology-driven industries:

Civil aviation. The United States is strong in this industry, with only one major, heavily subsidized competitor, Airbus.

Biotechnology. The United States is also a leader in this industry in which "the sky is the limit" and new discoveries are being made every month. Biology-based nanotechnology may someday be able to manufacture one atom at a time from locally available atoms. The biotechnology industry represents the convergence of several technologies, including computers, telecommunications, genetics, and micromachinery.

New materials. Steel, aluminum, plastics, and composites are current examples of new materials, and important new materials are still to come.

Microelectronics. The United States is the leading producer of high-value-added chips. Japan, which has focused on commodity chips, must now compete with the People's Republic of China, Indonesia, Korea, and others in the commodity chip market.

Computers and software. Because the United States has nearly 50 percent of the installed computer capacity in the world, it is in a strong position in the computer and software industry. Japan is second, with about 10 percent of installed capacity. The business world is interested in computers, but computers take a while to internalize, and the first generation of users may actually be less productive. This is in contrast to the Xerox machine, which changed behaviors and roles (e.g., the role of the secretary) very quickly by eliminating the need for carbon copies. Recent OECD data indicate that the United States is first in the growth of the service sector, which shows that U.S. business enterprises are effectively digesting new computer technologies.

Telecommunications. The United States is moving rapidly toward a high-speed, broadband, interactive information superhighway, provided it is not hijacked by government regulations. Telecommunications has had a real impact on everything from education (making home schooling and many other options possible) to decisions about location (largely eliminating the importance of distance).

Robots and machine tools. Robots and machine tools is the one major sector in which the United States is not even on the radar screen, although there are signs that it is making a comeback. Many other "comeback industries" in the United States, including heavy motorcycles (Harley Davidson), that were about to go out of business are now world leaders. Photocopiers (Xerox) is a resurgent industry in which developments in digital high-definition television have leapfrogged the Japanese. The Hewlett Packard inkjet printer also leapfrogged old technologies produced by Asian competitors.

Burgess noted that even though it is difficult to predict the importance of specific technologies, the United States is strong in six of the seven technology-driven industries.

Burgess then went on to discuss important historical changes that resulted from new ideas. For example, Jesus' ideas of love and hope changed the world, and Einstein's idea of relativity fundamentally changed perceptions. These pure ideas were not technology driven or coupled with experimental science.

Burgess called Christopher Columbus and Martin Luther the two most important examples of men whose ideas, coupled with technology, have changed the way we think. Christopher Columbus had a "big idea," namely that you could sail west to go east. His voyages were made possible by technological advancements, namely the astrolab, which made it possible to locate the latitude of a sailing vessel on the globe, and the caravel, which made it possible for ships to sail into the wind. Political factors were also important. The fall of Constantinople to the Muslims forced Western Europeans to find an alternative route to the East. Burgess noted that all of the major figures in the Renaissance were less than 25 years old when Columbus came back from the New World, except for Leonardo da Vinci, who was 40 but who did his most important work after that.

Twenty-five years after Columbus, Martin Luther expounded the idea of the priesthood of all believers in his 99 Theses. Burgess noted that Luther's idea was made possible by the invention of the Gutenberg press 62 years earlier. Within 10 years of that invention, the Bible had been translated into 10 languages, including German and French, which enabled people to read the Bible themselves.

New technologies have unleashed powerful social and economic forces that have had an enormous impact on our lives. Dramatic changes have been made in the workplace as the result of telecommunications technology. The number of temporary employees has increased, and freelance professionals (nomads) can move from job to job, enabling companies to adapt to a "project management" approach. The increasingly mobile workforce is possible because of "telecomputing" technology (the combination of computers and telecommunications).

Burgess believes these are positive changes. Nomads, for example, who continue to learn as they provide advice, counsel, and other services and then move on, are conduits for the rapid spread of ideas and the rapid diffusion of technology throughout the country, which has contributed to rapid innovation. Burgess believes that in the long run everyone will benefit from this trend.

Burgess also believes that telecommunications have enabled the just-in-time (JIT) office. Offices are becoming smaller, and the average office area, per professional, has dropped from 330 square feet to 110 square feet in many business and professional enterprises that are taking full advantage of new communications technologies. This change will have a profound effect on the real estate market. In addition, the spread of telecomputing technologies has had a profound effect on lifestyles. Compared to 1989, twice as many people work at home. A dramatic example is the phenomenon of "Lone Eagles," freelance professionals (knowledge workers) who have moved to small cities and towns and rural areas, especially in the Great Plains and Rocky Mountain region. This trend has been enabled by faxes, modems, express mail, and other transportation and telecomputing-based services and is creating a rural renaissance in the United States and a new way of thinking about economic development.

GEO-ECONOMICS OF 2020: THE GLOBAL MACROECONOMIC BACKGROUND

Edward Leamer

University of California at Los Angeles

The subject of Edward Leamer's presentation was the effect of technology on the standard of living. He pointed out that since the 1970s, real wages in the United States have declined, the inequality in incomes has increased, and the gap is growing (see Figure A-1). Compensation rates for the lowest 20 percent have fallen, which has had a dramatic effect on the political scene. The forces driving inequalities in income in the United States are education, immigration, globalization, and technology. According to Leamer, inequality in incomes has increased as the quality of a high school education has deteriorated. Immigration, predominantly low-skilled workers from Mexico and Central America, has increased the supply of low-skilled workers and lowered wages.

Leamer believes that globalization has increased the fluidity of products and financial capital. Manufactured products tend to level wages because they represent durable and transportable "stores" of human-value input. As more and more previously isolated economies, such as China, India, and Brazil, increase their trade with industrialized markets, huge numbers of unskilled workers enter the manufacturing labor force in which U.S. laborers must compete. If wage levels were equalized globally, they would equal $2/hour for all countries. Leamer be-

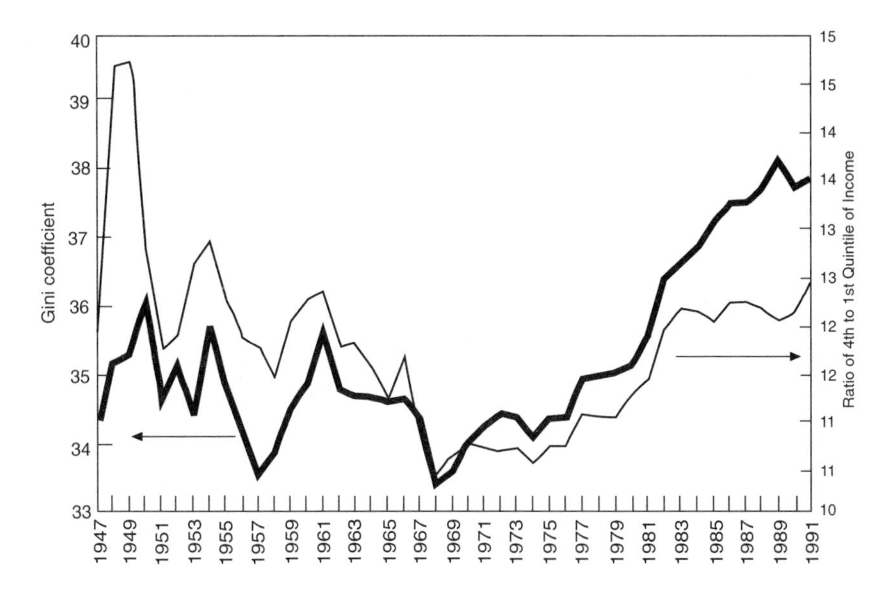

FIGURE A-1 Measures of inequality in U.S. incomes. Gini coefficient is a measure of income equality that ranges from 0 percent (indicating perfect equality) to 100 percent (indicating perfect inequality). Source: U.S. Census Bureau, Current Population Reports.

lieves that global wage leveling has increased inequalities in incomes in the United States. If low-cost, third-world labor can be substituted for high-cost U.S. labor, wages for low-skilled U.S. jobs will be limited or might even decrease. At the same time wages for more-educated workers with higher skills will increase. Industries that require substantial numbers of low-skilled laborers (e.g., manufacturers of shoes and apparel; see Figure A-2) are moving their operations to countries with low labor costs.

Leamer pointed out that new technologies can increase or decrease inequality in incomes. Some technologies, such as the forklift, increase the output of the operator in such a way that the physical capabilities of operators are equalized, because with a little bit of training, everyone can lift the same load and be paid the same amount. Therefore, "forklift" technologies tend to equalize incomes. Technologies that amplify the execution of tasks, such as the microphone, television, and CDs, enable single, talented individuals to reach much larger audiences than before. These "microphone" technologies create high rates of compensation and tend to increase inequality in incomes, which cannot be undone by education. Leamer asked workshop participants to consider whether the computer is a forklift or a microphone technology.

Despite advances in transportation and communications, Leamer asserted that proximity to major markets is still a principal factor in determining a region's per capita income (see Figures A-3 and A-4). He defined "law of gravity in trade"

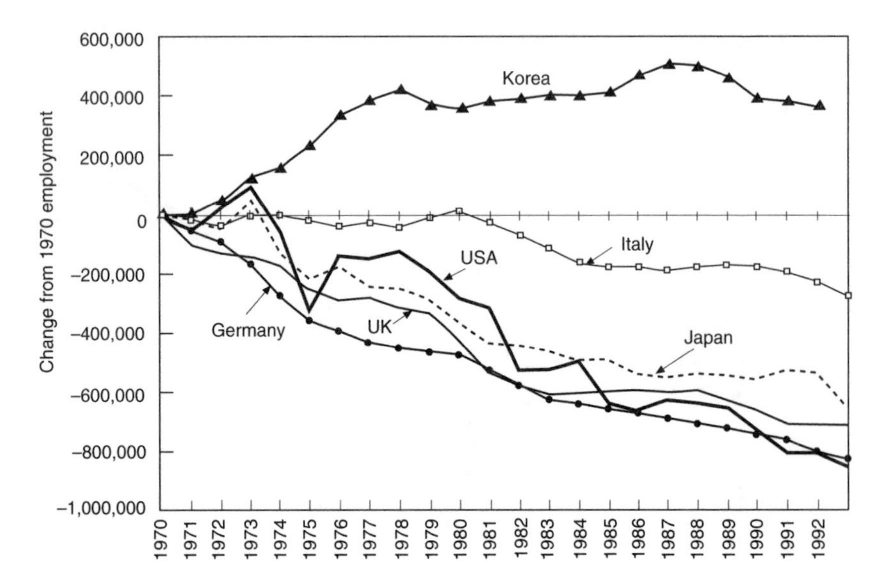

FIGURE A-2 Employment in the textiles, apparel, footwear, and leather industries, 1970 to 1993.

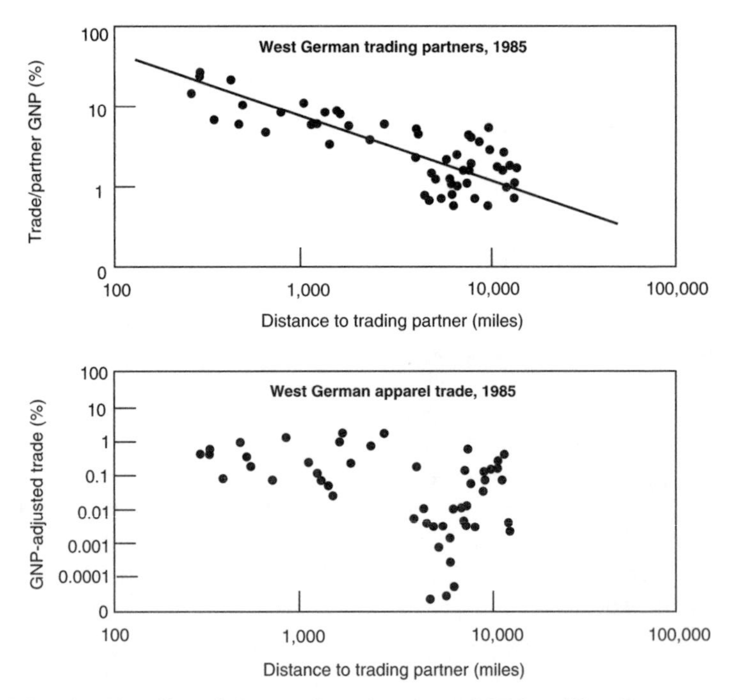

FIGURE A-3 The effect of distance from the center of GDP on West German total trade and apparel trade.

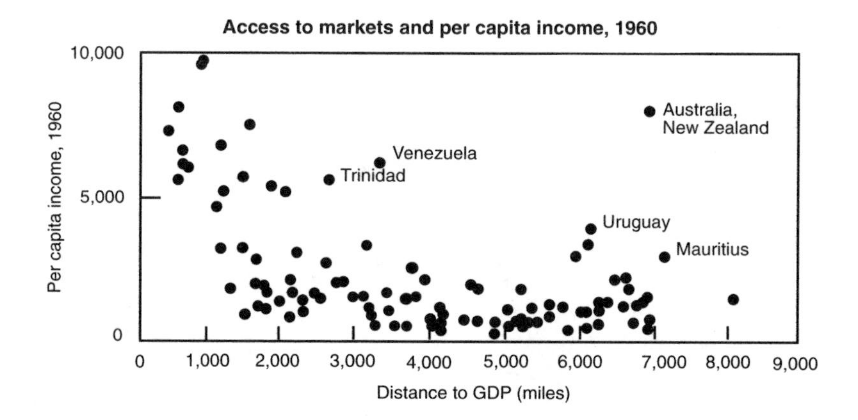

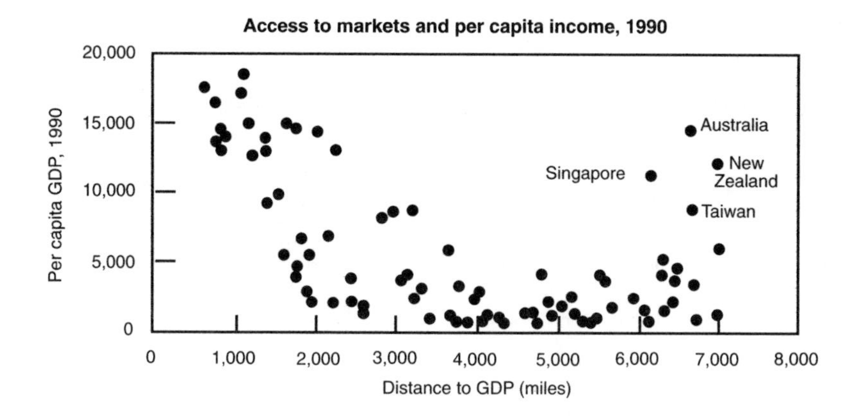

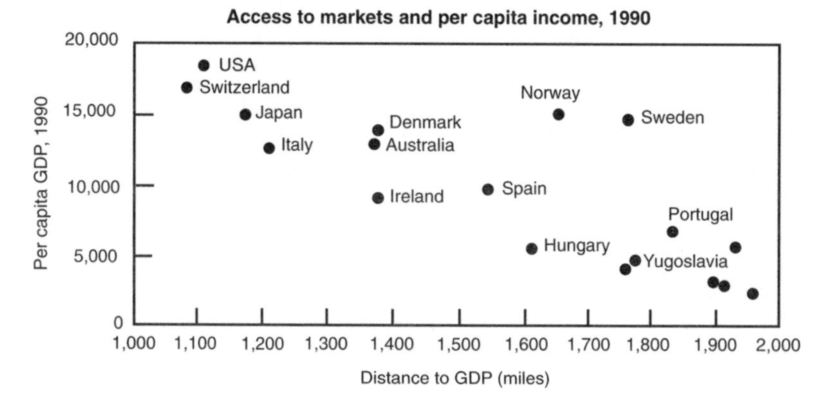

FIGURE A-4 Geographical clustering of high-income countries.

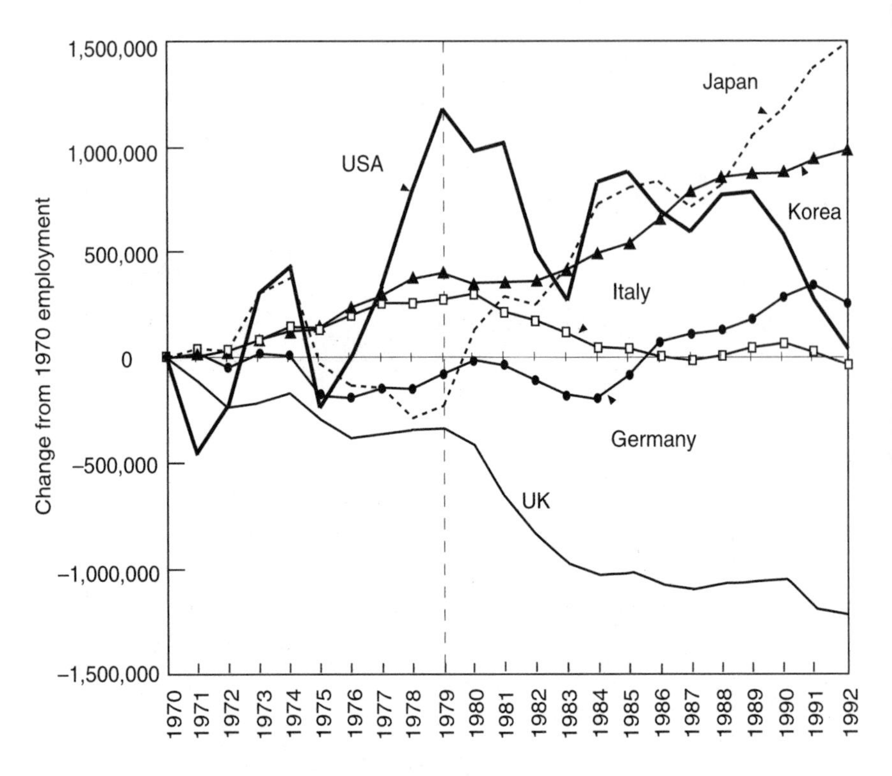

FIGURE A-5 Employment in machinery and equipment, 1970 to 1993.

as *the product of the gross domestic products of two countries divided by the square root of the distance between them is equal to the trade between the two.* This relationship remained relatively unchanged from 1975 to 1990, except for the shipping of automobiles across the Pacific Ocean, an anomaly that is unlikely to continue. The existence of infrastructure (including transportation, communications, education, and financial markets) is also an important factor in maintaining a region's per capita income. Therefore, Leamer believes that investing in infrastructure and education will minimize the negative effects of globalization on the United States.

The shift to higher-value jobs (e.g., the production of machinery and equipment) could be distorted by shifts in investment accounts (as opposed to trading accounts) that affect exchange rates and, consequently, prices. This is illustrated by the loss of U.S. jobs in machinery and equipment (Figure A-5) between 1979 and 1992 (particularly between 1989 and 1992), which was driven by a large increase in Japanese investments in the United States. This situation has largely corrected itself as the yen/dollar exchange rates have readjusted.

ENVIRONMENTAL CONSIDERATIONS IN MANUFACTURING

Paul Sheng

University of California at Berkeley

Braden Allenby

AT&T, Murray Hill, New Jersey

Paul Sheng began his presentation by stating that environmental impact is the product of three factors: population, which is growing; the wealth per unit of population, which is also growing; and the environmental impact per unit of wealth, which may or may not be growing. Sheng raised the question of whether or not this third factor could be used to compensate for the growth in the other two.

Sheng described three principal approaches to addressing the relationship between technology and the environment: remediation, compliance, and industrial ecology (see Table A-1). Remediation is a command-and-control approach that focuses on the past. The goal of remediation is to reduce local risk; environmental costs are treated as overhead. The interval between the generation of waste and remediation is very long, which creates difficulties with design and accountability. Compliance is another command-and-control approach that focuses both on the past and the present. Government agencies set an environmental standard for industry to meet; if industry meets that standard, government often raises it. Compliance is similar to remediation in many ways in that it also focuses on reducing local risk and treats environmental costs as overhead. The third approach is industrial ecology, or design for the environment. This new approach, which is currently gaining acceptance, represents a strategic and integral attempt to prevent or minimize adverse environmental impact. Industrial ecology was the subject of Sheng's presentation.

Industrial ecologists approach industrial systems the same way scientists approach biological systems. Industrial ecology is based on the entire life cycle of a component. In automotive technology, for example, the industrial ecologist approach would consider the following factors: the automotive subsystems (e.g., engine) and their effects on the environment, from resource extraction to consumer use; the process of automobile manufacturing, reuse, and recycling; infrastructure technologies, including the technologies needed to maintain bridges, roads, and gasoline stations, without which the main product could not function; and social structures, such as residential living patterns.

According to Sheng, sustainable development will require an industrial ecology infrastructure, which includes implementation initiatives, such as materials models and databases to support the determination of environmental impact; a research agenda; and comprehensive risk assessment and prioritization. Achiev-

TABLE A-1 Three Principal Organizational Approaches to Technology and the Environment

Approach	Time Focus	Focus of Activity	Endpoint	Relation of Environment to Economic Activity	Underlying Conceptual Model	Disciplinary Approach
Remediation	past	individual site, media, substance	reduce local anthropocentric risk	overhead	command-and-control intervention in simple systems	toxicology and environmental science; reductionist
Compliance	present/ past	individual site, media, substance	reduce local anthropocentric risk	overhead	command-and-control intervention in simple systems	toxicology and environmental science; reductionist
Industrial Ecology/ Design for the Environment	present/ future	materials, products, services, operations over life cycle	global sustainability	strategic and integral	guided evolution of complex systems	physical, biological, and social sciences; law and economics; technology and engineering; highly integrative

ing sustainable development would also require implementation of environmental accounting, as well as initiatives by various industry sectors, such as designing products for the environment, practicing sustainable agriculture and forests, and adopting sustainable energy systems.

Sheng described a number of factors behind environmental issues:

- **Emerging standards for managing product life cycles,** such as ISO 14000, British Standard 7750, EMAS, and Energy Star.
- **A growing consumer preference for "green" products,** such as products certified by Blue Angel and Green Cross. The green movement is strong in Europe and will probably become stronger in the United States.
- **The internalization of environmental costs** for the abatement and disposal of wastes into production costs.
- **Product "take-back" initiatives governing end-of-life,** initiatives that have been stalled in the European Union but are still a potent force. Instead of buying products, consumers will take out long-term (lifetime) leases on them.
- **Broader extension of the total quality management (TQM) movement,** which some expert believe has reached a point of diminishing returns. Total quality environmental management (TQEM) considers a broader context that includes environmental considerations.
- **Globalization and disintegration of manufacturing supply chains.** The question is whether ownership of the intellectual content of a design entails ownership of the environmental problems that ensue.

An environmentally sensitive view of manufacturing would consider waste and recycled materials from each step of the conversion process as "raw material" for some other process, (i.e., an extended supply chain). Product and process parameters would be mapped to waste groups, and process maps would be linked to supply-chain maps. Sheng believes that environmental management is a good example of distributed information and that an integrated solution to environmental management problems can be facilitated through the Internet.

Sheng listed the following emerging issues in the integration of environmental considerations and manufacturing:

- the development of materials databases and generally accepted techniques
- the internalization of the costs/benefits of environmental activities (activity-based management)
- the development of environment-based performance metrics
- the integration of environmental factors into supplier relationships (transaction cost economics)
- the integration of environmentally friendly designs with existing infrastructures for concurrent engineering and design-for-manufacturing

- the active management of energy consumption, especially for information systems
- the rigorous design and optimization of end-of-life processes and systems, (e.g., the ability to dismantle cars as fast as they are assembled)
- the modification of the definition of a supply chain (industrial symbiosis) to include the end-uses of byproducts and waste

REENGINEERING THROUGH FRACTAL STRUCTURES

Wilfried Sihn

Fraunhofer Institute for Manufacturing Engineering and Automation, Stuttgart, Germany

Wilfried Sihn's talk was divided into two sections. In the first section, he described the re-engineering of German corporations using a so-called fractal structure. In the second section, he described his vision of the future competitive environment in Europe.

Sihn believes that the following factors affect corporate success: a culture of innovation, globalization, organization optimization, location safeguarding, diversification, and customer orientation. The successful corporate culture is characterized by cost management, employee orientation, process orientation, and production depth. In order to survive, companies must change their values from centralization to decentralization; mistrust to trust; power to communication; specialization to flexibility; determinism to chaos; and company tradition to company culture (see Table A-2).

TABLE A-2 Necessary Changes in Company Values

Old Values	New Values
"Centralism"	"Decentralism"
Mistrust	Trust
Outside control	Self-supervision
Division of labor	Work enrichment
Individual performance	Team performance
Output	Quality performance
Power	Communication
Notification	Information
Position	Executive responsibility
Hierarchy	Process orientation
Company tradition	Company culture
Specialization	Flexibility
Line	Network
Determinism	Chaos
Training	Motivation
Job orientation	Relation orientation

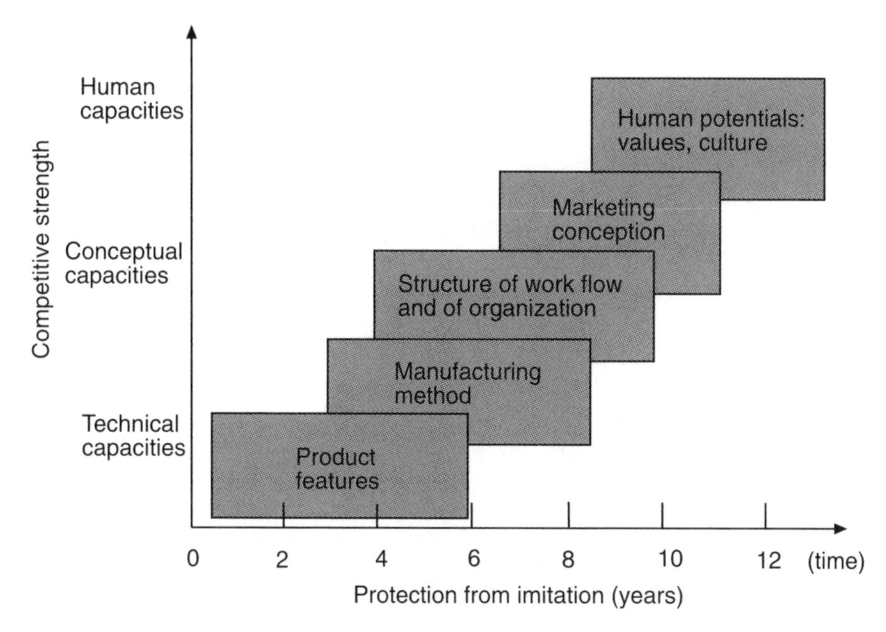

FIGURE A-6 Conceptualization of how lasting competitive advantages are attained through values and culture.

Sihn defined a fractal company as a business organizational structure with many of the properties of fractals: self-similarity, self-optimization, and self-organization. The Fraunhofer Institutes in Stuttgart and Magdeberg have used these concepts to develop a methodology for thinking about and implementing open organizational systems composed of small, semi-autonomous work units, or fractals. To date, this methodology has been used to restructure nearly 200 companies and has demonstrated improvements in business results.

In addition to the characteristics listed above, the fractal company places a heavy emphasis on the value of employees. One core axiom of a fractal company is competitive strength and the lasting competitive advantages of safeguarding a company's human capital (see Figure A-6). Leadership in a fractal company is not top-down. Instead, development teams bring employees on board. The fractal company includes six levels in a business organization: cultural, strategic, socio-psychological, economic/financial, information, and process and material flow levels.

In the second part of Sihn's talk, he described a vision of the competitive environment of the future. The key elements of this environment are listed below:

- The European Union has become a reality, and national borders have been dropped. Europe now acts as a unified whole in global markets with a single, very hard currency.

- China, India, and Russia have become economic world powers. The "tiger states" (Southeast Asia) have formed a solid union.
- Cost differences among world powers have narrowed to less than 100 percent.
- Europe, as well as the United States, is concentrating on high technology products and "intelligent" services.
- The power of labor unions has diminished.
- The social welfare state has become an unaffordable luxury, and the emphasis is again on achievement.
- Competition takes place in "electronic markets."
- Ecology (saving the planet) has taken priority over the economy.

Sihn predicts that, in this competitive environment, a few global enterprises will dominate international markets, leaving smaller companies to pursue regional and technological niche strategies. The European economic structure will be dominated by gigantic groups surrounded by many highly flexible and dynamic small businesses. European labor laws will permit highly flexible working conditions and remuneration. Companies will operate 24 hours a day, seven days a week. Product structures, assembly modes, and product delivery will be different than they are today. Business enterprises will be paperless, and networking will be worldwide.

According to Sihn, the major challenges faced by manufacturing enterprises in this environment will be the implementation of flexible, temporary cooperation models for virtual enterprises; knowledge management; value engineering; the creation of a culture of innovation; globalization; changing leadership strategies from confrontation to motivation and cooperation; resource shortages; competition in time; and competition in competence and cost. Sihn believes that the technological developments required to remain competitive in this environment will include recyclable materials; new ceramic, metal, synthetic, and biological materials; and multifunctional materials. Technologies that minimize the number of components and replace mechanical systems with electronic systems will also be necessary. Finally, key technologies, such as genetic engineering, environmental technology, semiconductor technology, mechatronics, and microsystem technology, will have to be developed.

THE INFORMATIONAL INFRASTRUCTURE OF 2020

H.T. Goranson

Sirius Beta, Virginia Beach, Virginia

H.T. Goranson began his presentation by discussing the role of technology in manufacturing. He pointed out that there are two types of technologies, "push" technologies and "pull" technologies. Push technologies appear unexpectedly and bring about profound changes in society by way of new types of products (cars, phones, televisions, computers). Only much later do these technologies affect the way manufacturing is done. Pull technologies, on the other hand, are information technologies that enable companies to work around difficult problems. To predict the technologies manufacturing enterprises will need in 2020, Goranson believes we must first determine the problems businesses will have to solve.

Goranson next discussed the relationship between manufacturing, collaboration, and technology. He believes that commerce, which is even older than government, is the basis of societal collaboration. In Goranson's opinion, the manufacturing enterprise is at the heart of collaboration, all collaborative processes are essentially about information, and collaboration is technology dependent. Goranson described eight future conditions or megatrends.

Brand loyalty plus. People already identify certain brands of products with their lifestyle and ethnic or group identity. This trend will continue.

Megawealth generators. As a result of brand name loyalty and other factors, a few brand name manufacturers will become powerful collectors of wealth. However, they will do less and less of the actual manufacturing, which will occur further down the supplier chains. One group will do marketing, another will do investing, and a third, the supplier base, will do the manufacturing.

Third, dynamic class. Markets and innovations will be highly dynamic. The number of "have-nots" will grow, and a new class of "used-to-haves" will emerge. The used-to-haves will be educated, motivated people who were "haves" but have become have-nots.

Wealth by ability to change. Wealth will be granted by the investor community. Immediate wealth (profitability) will be eclipsed by estimates of how profitable an individual or company is likely to be in the future. Wealth will be determined by impressions of an individual or company's ability to stay ahead of the power curve (i.e., by its agility). The ability of investors to maintain wealth will be based on how well they manage the supplier base.

Products as strategic weapons. Delighting the customer will become less important than "using" products as competitive weapons (the Microsoft/NBC model). This change has already taken place in the movie industry.

Lifetime product marketing. The social identification of the product will be leveraged for after-sale sales (e.g., lifetime improvements in autos). Today, many manufacturing enterprises only engineer, manufacture, and sell a product. Companies of the future will keep profiting by continuing to upgrade their products after sales.

New social roles for commerce. More social services will be performed by commercial entities under the primary investors (megaprimes), who will be less concerned with national issues. Social services will be associated with brand identification (e.g., the reinvention of insurance companies), and the role of civil responsibility (e.g., the role of haves in the supply chain) will be redefined. Suppliers will be invited to buy into health benefit pools.

Integration as the goal of research. Primary investors who can integrate processes against the soft market context, leaving the innovation of processes to supplier partners, will reap massive payoffs. The focus of research will shift from new development to integration for new markets.

Goranson believes that these megatrends will generate certain technological needs. He believes that manufacturing enterprises lack the tools to manage complexity, abstraction/aggregation, and "soft" (social/cultural) dynamics. A key information technology for the future will be the ability to identify product needs in soft contexts (this capability already exists for cars, shoes, food, and entertainment). Companies must be able to identify customer reactions to products before they appear on the market and create a demand for their products. Goranson believes that this can only be done with soft modeling. Companies must understand highly complex combinations of products and product factors; manage combinations of suppliers and processes to meet identified needs and extrapolate new possibilities; and optimize their operations to meet not only current needs, but also future needs. Companies will need these capabilities for dozens of products and millions of suppliers.

For these soft models to work, they must be deep in terms of formal mathematics. Things will change too fast to rely on intuition. However, the models must also be expressible in concepts that consumers and investors can understand. Soft models must also operate in such a way as to enhance the national good because the market can not be relied on to address these issues. Social metrics must, therefore, be incorporated into the models. Investments in technologies to meet business demands may differ from investments the nation would make to improve or maintain public health.

Goranson outlined a number of problems with existing approaches: product models are not tied to process models; businesses exploit social and cultural phenomena without the tools to evaluate the complexity of product combinations, the softness of projected demand, or the consequences of their actions; there is no formal modeling technology for soft dynamics; technology today tends to create homogeneity rather than diversity in the supply chain; there is no analysis-to-control linkage in our technology foundations; the complexity of the infrastructure is growing faster than the complexity of the enterprise.

Goranson suggested three grand modeling challenges for the research agenda of the future: models of soft phenomena, such as social and cultural dynamics and associated strategic goals; small, specialized groups that can use whatever processes, analyses, tools, or representations they desire without constraint, yet can be part of large, diverse enterprises; and models that lead to automatic binding and governing mechanisms that enable business aggregations to evolve automatically. Ideally, process modeling/knowledge-representation science could be combined with programming/natural languages to produce a language that can describe, explore analytically, and control complex soft systems. This language would have a computable internal representation (possibly based on multi-agent systems components) and multiple views, including spatial visualization vocabularies (possibly based on topographical manipulation).

LOOKING OUTSIDE THE BOX BUSINESS PRACTICES FOR INDUSTRIAL COMPETITIVENESS IN 2020

Rick Dove

Paradigm Shift International, Oakland, California

Rick Dove began his highly visual presentation by stating that we can't imagine 2020 because we can't look "outside the box" to foresee revolutionary developments. Then he described some provocative ideas about what manufacturing in the future might be like: autonomous self-organizing systems will be common; people will have to cope with rapidly changing technology; businesses will generate value in blitzkriegs; the people in power will be today's 10-year-olds, who will be practically omnipotent and immortal; technology will change so quickly that luck will be more important than strategy, as we know those concepts today; and business value will only be short term.

Dove believes that within five years, we will no longer recognize the business world. Laser sintering of useful metals, as well as atomic construction, will be possible. An Internet satellite grid 200 miles above the earth will be in place. Virtual reality will be used as a cooperative work space, and employees and customers will be plugged-in cyber people.

Dove believes that the pace of change, both technical and cultural, will continue to accelerate over the next 25 years. In Japan, a computer generated pop star, Kyoko Date, already has hit songs, and, because of MTV, African children living in tents want Nike shoes. The speed and intensity of life will be enhanced by electronic, medical, and other technologies, including drugs that can enhance learning and problem-solving capacities. He believes that in 25 years, the 10-year-olds of today:

- will have drug- and genetically-enhanced mental powers
- will be economically pulled, not driven
- will be intellectually motivated
- will want enough money to buy a life
- will have the option of living forever
- (some) will be Goldfinger-type criminals
- will be able to live anywhere without regard for international boundaries

According to Dove, we can't afford to learn history anymore. Instead, we need to learn how to learn and what insight means and that a lot of good strategy can be learned from playing "Doom" (an electronic video game), although perhaps it is dangerous to teach people that they can "save and reload." In the future, people will be "wired for sight and sound" and will be able to get information simply by asking for it. Virtual reality will be a national utility, and entertainment will be immersive and rewarding.

In his presentation, Dove conceptualized business of the future as a collection of nonlinear systems composed of interacting, independent modules. He suggested that companies will keep expanding until they try to tackle projects that are beyond their competency (e.g., DEC and Apple) and that management consulting in 2020 will include the service of dismantling companies while they are still net positive (the Kevorkian Group). Long-term commitments to companies and countries, he said, will give way to short-term opportunistic relationships that may be complex and far flung. Labor unions, as we know them, will become weak, and there will be a return to guilds and an emphasis on continuing education. Dove believes that women will be dominant in business of the future because, genetically, they have better social skills.

Dove predicted that products will emerge and disappear rapidly, emulating the practices of the fashion and entertainment industries. Many companies will abandon their primary commitments to self-preservation and growth, changing fundamentally to opportunity-specific enterprises, much like the transition in the film industry from large studios to independent ventures that assemble and disassemble with the product life cycle. This implies that the assembly/disassembly process must be made much easier. New control strategies (e.g., autonomous agents) promise the ability to control complex interactive systems with a few simple rules. The important principle, according to Dove, is "united we fall, divided we stand" because tightly coupled systems are brittle and move slowly, while loosely coupled systems are flexible and can change quickly.

Dove suggested that in the year 2020, there will be people practices, rather then business practices, and that there will really be no time for businesses to practice at all; they will "just do it." Trends toward less constrained, more autonomous units will present challenges to ensuring ethical, or even "legal," behavior as autonomous, flexible, and unconstrained business units explore and test boundaries and borderlines in all directions. Dove believes that the innate competitiveness of people will preclude trust-based business relationships in the foreseeable future.

SUMMARIES OF GROUP DISCUSSIONS

The group assignments for the first day of the workshop are listed below. The first name is the committee member who acted as facilitator for the group, and the names in italics are the spokespersons who presented the results:

Group 1: Ann Majchrzak, *Nathan Cloud*, David Hardt, Louis Kiefer, Howard Kuhn, Woody Noxon, Paul Sheng, Mauro Walker
Group 2: Barbara Fossum, Debra Amidon, *Thomas Crumm*, Robert Hocken, Edward Leamer, Leo Plonsky, Wilfried Sihn, *Brian Turner*
Group 3: David Hagen, John Decaire, Bill Kay, *Mike McEvoy*, F. Stan Settles, Patricia Whitman, John Bollinger
Group 4: Gordon Forward, Steven Bomba, *Rick Dove*, Richard Jarman, Rakesh Mahajan, Eugene Wong
Group 5: Donald Frey, Richard Kegg, David Miska, *Richard Neal*, James Solberg, H.T. Goranson
Group 6: Lawrence Rhoades, Charles Carter, William Hanson, M. Eugene Merchant, Richard Morley, Heinz Schmitt, *Kathryn Whiting*, George Hazelrigg

Each group was asked to consider the following questions:

1. What are the most important challenges the manufacturing industry must address to compete successfully in 2020?
2. How will manufacturing be done in 2020?

The following sections contain the responses of the discussion groups.

GROUP ONE

Question 1: Manufacturing Challenges in 2020

Challenges for the manufacturing enterprise in 2020 will include anticipating and defining the concept of adding value in a much more dynamic market; providing satisfying challenges for people; and creating dynamic organizational constructs that can integrate multiple points of view, such as local vs. global, employee vs. employer, entrepreneurial vs. company, nationalistic vs. profit boundaries, and good and bad aspects of a nomadic workforce. The challenges in the areas of growth, management of the global supply chain, and the integration of multiple perspectives are listed below:

Growth

- defining products (blurring of the distinction between services and products)
- determining who should be involved in setting business strategies (stakeholders)
- defining value and how manufacturing enterprises should provide value to customers
- determining and maintaining the core competency of manufacturing
- establishing metrics for long-term growth

Management of global supply chains

- managing the logistics of global supply and capacity from region to region
- integrating the supply chain and product concepts
- accommodating changing local markets (instead of shipping)
- rapidly creating and dissolving supply chains

Integration of multiple perspectives

- rationalizing functional, resource, and organizational perspectives
- integrating knowledge and skills in manufacturing enterprises
- optimizing the relationship of employees to manufacturing enterprises
- determining roles for political entities in the transition from local to global enterprises
- preserving political stability

Question 2: Manufacturing in 2020

The group agreed that manufacturing in 2020 will be based on innovation and the development of new products, rather than on filling market voids. Manufacturing will include the entire supply chain, i.e., marketing, distribution, design, and the in-home manufacturing and assembly of goods.

There will be more than one type of manufacturing organization in 2020. Manufacturing organizations will vary in virtuality and loyalty (see Figure A-7). Virtuality will vary from mega-companies (fractal organizations) to virtual networks, and the loyalty of the employees will vary from a loyal core of integrators and professionals (with the range of skills necessary to a particular business) to free agents. A company's core competencies will be focused on expert knowledge of the business, with a lot of outsourcing. Employees will have incentives to go wherever they want or wherever they can add value.

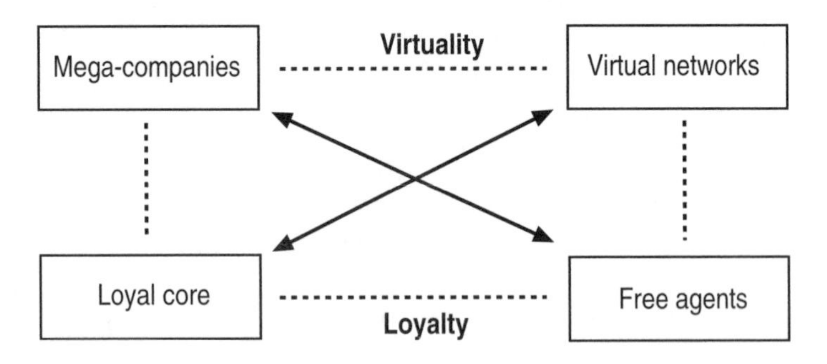

FIGURE A-7 Schematic illustration of how manufacturing organizations in 2020 will vary in virtuality and loyalty.

Elements of successful manufacturing in 2020 are listed below:

- processes that are understood and consistently executed
- new materials
- environmentally friendly products and processes
- mass customization along with mass production
- local manufacturing
- machines that are easy to operate and repair
- networked/holonic/virtual work organizations along with extended mega-corporations
- companies with access to all educational systems
- *in situ* sales, distribution, and manufacturing
- knowledge management
- constant and rapid product innovation
- reconfigurable/reprogrammable factories
- global enterprises

GROUP TWO

Question 1: Challenges for Manufacturing in 2020

The group first addressed the question of the most important challenges that the manufacturing industry must consider to compete successfully in 2020. The challenges identified were grouped into four areas, (1) managing knowledge, information, and communications, (2) operating global enterprises, (3) sustaining the manufacturing infrastructure, and (4) managing change.

Managing knowledge, information, and communications. The physical aspect of managing knowledge is exemplified in the ability to unbundle physical products from information about how to use them. The human aspects of managing knowledge, information, and communications include retraining workers for advanced processes and understanding cultural change (for example, the increase in life expectancy, the decrease in the number of children per family). Other specific factors that were identified in this category include: capturing and implementing new technology, cross-boundary processes, protection of intellectual property, changing rules, funding knowledge development, and funding of entrepreneurial efforts.

Operating global enterprises. Issues related to the globalization of manufacturing enterprises include accommodating multiculturalism (e.g., a multicultural workforce), choosing the right market for a product, doing business in various economic systems, and redistributing technical competence and wealth (e.g., understanding the dynamics of third-world countries and assessing the impact of their growing capabilities). Enterprises will have to determine which technologies and strategies they should pursue in order to address these issues.

Sustaining the manufacturing infrastructure. Sustaining the manufacturing infrastructure includes creating new components of the infrastructure, such as components related to energy and natural resource use, the environment, education, and transportation, because current components will have to be replaced by 2020.

Managing change. Managing change will require an understanding of social and cultural changes (longevity, family size, demographics), an understanding of systems dynamics, thinking and acting in new ways (collaborating, converting dilemmas to opportunities ["and" versus "either/or"]). Managing change will also require recognizing that industrial changes must parallel and accommodate societal changes.

Question 2: Manufacturing in 2020

Group 2 identified the following characteristics of manufacturing in the year 2020:

- Fundamental industrial processes (e.g., unit processes) will still be around and will not be much different. Milling and welding will not be obsolete.
- Products and services will be increasingly produced in multiple configurations of alliances, some of which will even include consumers.
- Networks of flexible entrepreneurial sites will replace large, more rigid, central sites. Examples from the past two decades are the mini steel mills.
- Products that are built-to-order, instead of inventoried, will create pressures to reduce scale and locate sales and services near manufacturing sites. Selecting products from store shelves will increasingly be replaced by factory-direct orders and special delivery options via communication links.
- Industry will produce more highly customized, high value products. Bicycles, for example, will be made to fit not only the body size, but also to accommodate the physical condition of the rider.
- New developments in biotechnology and nanotechnology will create whole new industries and industrial alliances.

The following scenarios were also discussed:

- More emphasis will be placed on remanufacturing and extending the life of products.
- Humans and machines will work together more closely.
- The workforce will be more educated, more capable, and more responsible.
- Manufacturing will be done in networks of organizations, "global production networks."
- We will continue to mix high-volume mass production with low-volume and high-volume customization.

- Manufacturing could disappear by 2020.
- The minimum replaceable component of equipment will be increasingly complex and expensive.
- Manufacturing may be done in outer space.
- Economic security and job security will continue to decline causing inequality in income to increase substantially.
- Organizations will drift in and out of alliances.

GROUP THREE

Question 1: Challenges for Manufacturing in 2020

The discussions of this group centered on the premise that manufacturing companies in 2020 will have to be extremely efficient, highly competitive, extremely agile, and extremely responsive to changing customer requirements and competitive conditions. Companies will have to optimize their performance by fully utilizing diverse global human resources. Based on this premise, the group identified eight critical challenges: managing the enterprise as a system; educating the workforce; accommodating cultural diversity; managing knowledge; managing environmental impacts; adapting to social instabilities; sustaining customer relationships; and managing innovation.

Managing the enterprise as a system. All highly distributed, global companies that are quick and responsive will manage their enterprises as systems. Savings from low-cost labor are likely to be offset by an increase in time materials or products remain in inventory. Companies that operate as systems will be able to find the most qualified people and allow them to work together to reach common objectives.

Educating the workforce. Education, and the way we perceive it, will change by 2020. Because all global manufacturing enterprises in 2020 will have virtually unlimited access to educational systems worldwide, education will offer no competitive advantage. However, the effective utilization of human and educational resources will offer advantages. Educational challenges for future manufacturing enterprises will include maintaining a "fresh," competitive, up-to-date workforce and developing new competencies to stay competitive in a changing business climate.

Accommodating cultural diversity. Manufacturing enterprises in 2020 will require highly distributed decision making to deal with the fast-moving global business climate. Decision making, however, will have to be guided not only by the high level "systems view" described above, but also by shared values and operating principles. The challenge to future manufacturing enterprises will be to create and sustain shared values and principles and to make full use of the human resources that will be available globally.

Managing knowledge. Manufacturing enterprises in 2020 will have access to essentially all information relevant to their operations. The challenges to future manufacturing enterprises will be to transform the available information into useful knowledge for all relevant parts of the value chain and to manage knowledge assets in a way that yields competitive advantage. Some discussion participants described the concept of the "thin edge" as critical, time-perishable product and knowledge assets. Manufacturers will have to decide how much of their critical assets should be shared in a collaborative relationship.

Managing environmental impacts. The manufacturing climate of the future will be much more sensitive to environmental issues. The challenge to manufacturing enterprises in 2020 will be to take advantage proactively of environmental considerations rather than being driven by them.

Adapting to social instabilities. Social instability might be greater than it is today. Instabilities will arise from the differences between the haves and the have-nots.

Sustaining customer relationships. A critical challenge to manufacturing enterprises in 2020 will be to develop and retain customer loyalty. This will require sustaining customer relations and giving customers what they want, when they want it.

Managing innovation. Managing innovation to provide products or services that are valued by customers and that return profits and competitive advantage will be a key challenge to manufacturing enterprises. If a company does everything else right but does not innovate, it will still fail.

Question 2: Manufacturing in 2020

Participants identified the following characteristics of manufacturing in 2020.

- Customers will drive manufacturing to an effective production batch size of one for high value-added products. System flexibility will be required for undreamed of customization of products. Even today, medical equipment is produced in batches as small as 10.
- Global networking and collaborations will continue to expand, allowing companies to change their main products quickly. Companies are already striking relationships never before dreamed of. Knowledgeable engineers and the ability to manufacture locally will be the key to reaping the benefits of collaborative enterprises.
- No differentiation will be made between products and services. *Fortune* magazine quotes a Matsushita saying that service after sales is more important than assistance before sales because it is the way one wins permanent customers.
- Virtual corporations will integrate major products. For example, the Boeing 777 was characterized by one participant as "a large assembly of

precision milled parts flying in formation." This type of product integration is already happening today. A medical (intravenous solution) pump was recently designed in the UK. The fabrication tools included 143 molds fabricated on three continents—Europe, North America, and Asia—in 90 days. These tools, in turn, generated finished parts that were shipped to Singapore for final assembly.

- The interdependence of core partners in the value chain will be a barrier to establishing new collaborative partnerships because of the established business dependencies and intellectual property shared by the partners. Small manufacturing enterprises will have to be able to manage multiple partnerships.

GROUP FOUR

Question 1: Manufacturing Challenges for 2020

Some participants identified the following important challenges that will be faced by the manufacturing industry in 2020:

- attracting people to careers in manufacturing
- responding to catastrophic events, (e.g., economic collapse in China)
- developing employees who are skilled at knowledge work, which will become important in the factory environment and provide competitive advantage
- keeping employees up to date
- defining the government's role in manufacturing
- optimizing energy use

Question 2: Manufacturing in 2020

The group discussed how manufacturing would be done in 2020 and identified the following trends:

- Mass production and mass customization of products will be prevalent. But mass production will require reconfigurable factories. Despite the prevalence of mass customization, economies of scale will still have advantages. Some products will be generic and some will be custom-made. Products will be designed to be reconfigured and manufactured for longer lifetimes.
- Factories will be flexible and programmable so that they can manufacture a wide variety of products. Key issues in developing programmable factories will be determining the number of required workers and their required skills. Workers will have to be knowledge workers rather than machine feeders.

- Manufacturing enterprises will be dispersed networks of small manufacturing cells. Transportation and information networks will link the cells into productive agglomerates of networked systems. There will be no distinction between large companies with internal networks and networks of smaller, specialized companies (i.e., both types of organizations will be managed the same way).
- Interest in craftsmanship will be revived, enabled by commodity production resources. Custom-made furniture and other products for the home will be popular. The affluence in the United States and the ability of U.S. manufacturers to produce small quantities of products cost effectively will allow us to indulge these whims.
- Inexpensive, custom-made items will be enabled by an installed infrastructure of outsource manufacturing.
- Knowledge workers will be able to configure and reconfigure factories. Few, if any, workers will be "on the floor."
- Machines will be easy to operate and maintain.

GROUP FIVE

Group participants identified the following global crises that they believed could affect manufacturing enterprises in 2020:

- global energy crises, e.g., the depletion of fossil fuels
- social crises, e.g., war between the haves and the have-nots
- environmental crises, e.g., shortages of clean water
- monetary collapse
- global health crises, e.g., pandemics
- social upheavals caused by a radical shift to cost-effective manufacturing
- vulnerability of information, e.g., breaches of computer security

Question 1: Challenges for Manufacturing in 2020

The group discussed the challenges to manufacturing that would result from the crises described above, including changes in living conditions, education (with industrial involvement), changing the skill base to a knowledge base, establishing meaningful reward systems, and resolving cultural conflicts, defining communities, and changing the structure of companies. The key challenges to manufacturing in 2020 identified by discussion participants are listed below:

- partnering to compete and cooperate
- responding to increasing customer demands
- taking responsibility for adverse environmental impacts
- balancing workers' cultural needs with the ability to perform
- educating the "emerging workforce," e.g., accommodating the growing disparity between skilled and unskilled workers

Question 2: Manufacturing in 2020

The group next discussed what manufacturing would be like in 2020. The group discussed time as an organizing principle, distributed manufacturing, increased automation, and the merger of service and manufacturing. According to some participants, challenges to manufacturing enterprises in 2020 will include the development of upgradeable product platforms, the presumed necessity to surrender autonomy in favor of collaborations, and the protection of knowledge assets. Group participants identified the following important trends:

- Extended enterprises will be dominant. Major corporations will exist, but will produce very few parts. Instead, they will be brokers who design, assemble, and manage.
- Small businesses will continue to thrive. The businesses that can adapt will survive. Investment capital will be widely distributed and dispersed.
- "Presence" will be a corporate goal. As economies, such as Indonesia and China emerge, companies will want to have a local presence.
- The business environment will be a even more hard-nosed than it is today and dominated by the economic "bottom line."

Product realization will be quite different in 2020 in terms of responsiveness and affordability. Manufacturing processes will be well understood and consistently executed. The group identified the following three key characteristics of the manufacturing climate in 2020:

- Distance, national boundaries, financial differences, and information issues will no longer be barriers, but will be key factors for decision making.
- New materials (e.g., biotechnology and composites) and new production processes (e.g., molecular manufacturing) will present new opportunities and challenges.
- Energy and environmental responsibility will be integral to successful manufacturing.

GROUP SIX

The first point of discussion was to identify geopolitical events that would influence the manufacturing climate in 2020. Following an extensive discussion, participants advanced the following scenario, with ecology as the primary driver: A major ecological disaster will trigger a dramatic strengthening of the "U.N. Security Council/Group of Seven" community, which will establish a Super Power Federation with "teeth" (i.e., economic and military authority). This will lead to a dramatic reduction in terrorism, an increase in open trade, and strictly enforced global environmental standards. Some implications of this scenario are listed below:

- fewer asset-depletion economies
- increased importance of remanufacturing of products
- accelerated globalization of multinational enterprises as the political situations in third world nations become more stable
- continued domination of the production of high-technology weapons by superpowers (particularly the United States)

Question 1: Manufacturing Challenges in 2020

Many U.S.-owned and -controlled corporations have factories, partners, suppliers, and customers throughout the world. Foreign-owned U.S. factories is is one of the fastest growing industrial sectors of the U.S. industrial base. The most "liquid" stakeholders in a publicly traded company in the United States are the shareholders, many of whom own shares through an institutional investor and don't even know they are shareholders. Their commitment to the company is less reliable than a typical employee's. Small business owners have the longest term (and deepest) commitment to their companies. The implication is that companies seeking long-term commitments from core employees should compensate them with nonliquid equity. Other stakeholders include management, customers, suppliers, and communities (i.e., taxing authorities), all of whom depend on the company economically in some way. Providing products, reward systems, and organizational structures that more effectively resolve the conflicting interests of these stakeholders could dramatically reduce the waste and stress of intra-enterprise conflicts.

The challenges identified by group participants are summarized below:

- resolving the interests and conflicts of stakeholders
- protecting knowledge assets while increasing employee mobility
- attracting manufacturing employees by maintaining a high standard of living, providing education, and improving management
- changing from skill-based to knowledge-based enterprises
- making manufacturing more attractive as a career
- rewarding employees
- focusing on the future

Question 2: Manufacturing in 2020

Some group participants discussed the scenario of manufacturing performed by "holonic" systems of a core organization (e.g., Fortune 100 manufacturers) supported by a select group of networked and cooperatively agile partners (smaller suppliers). Small suppliers will compete to become and remain "members" of this team, surrendering *some* of their autonomy to the group (i.e., the core organization). The team will be reevaluated periodically, with purges and trades to ensure the most talented and well-balanced team composition.

The high cost of transportation and distribution will lead to distributed local, neighborhood, and personal factories. These distributed factories will access component design and machine control codes from design owners electronically and will construct components from powder materials that can be blended and alloyed to meet the specific requirements. Excess capacity in personal factories can be used to produce commodity products for the "MasterCard" market (i.e., an electronically networked "Dutch Auction" market that solicits component production to be delivered to the buyer by a target date and at target prices from any producer in the network).

Other possible trends for manufacturing in 2020 that were identified by some participants are listed below:

- Service and manufacturing will merge and products will be upgradeable.
- The contributions of individuals and teams in complex manufacturing operations will be accurately measured.
- Substantial time and effort will be devoted to planning for the future (i.e., establishing long-term objectives of the business).
- Time will be recognized as an important contributing factor to cost, particularly for products that become rapidly obsolete.
- Manufacturing will be concentrated in areas where smart, well educated people want to live (the implication is that maintaining attractive living conditions will protect U.S. manufacturing).
- Manufacturing infrastructure (i.e., transport, communications, education, and supplier base) will be more important than labor costs.
- Education will become much more efficient, utilizing "microphone" technologies (á la Edward Leamer) to extend the reach of talented teachers with the help of "Steven Spielberg" presentation technologies.
- Education will be highly valued and aggressively sought at all levels.
- The selection of managers will be less arbitrary and less artful.
- Companies will develop sophisticated strategies to protect and exploit their knowledge assets.

PART II

Manufacturing Technology in 2020

WORKFORCE ISSUES OF 2020

Brian Turner

Work and Technology Institute, Washington, D.C.

Brian Turner began his talk by describing the historical trajectory of change in the organization of the manufacturing enterprise from mass production to lean production to total quality production to agile/fractal production and, finally, to high performance production. The manufacturing sector is currently characterized by rapid changes and innovations, and this trend will continue. Innovations in the manufacturing sector have also affected the service economy, including flattened organizational hierarchies and empowered workforces; pervasive information and communication technologies; and pressures for better performance in terms of quality, cost effectiveness, and speed.

Turner described the workplace as a combination of technology, skills, and organization, all three of which are rapidly changing. Because people are the central, and most important, element in all three, the technological choices made today will affect not only quality of the workplace, but also the society in which we live. Technological choices can be approached from three perspectives: a technocentric perspective, in which social dimensions are ignored; a sociocentric perspective, in which technology is ignored; and an integrated perspective, which takes both into account.

Turner believes that three areas of technological development will be critical for the workplace in 2020. The first is the development of educational systems for teaching basic skills, such as math, science, and literacy; technical skills; high-performance skills, such as communication, problem solving, quality, and team operations; and continuous learning. These educational systems are necessary to increase the supply of skilled workers. The second technological area is the high-performance organization of work and production. This might include flattened and simplified hierarchies and active partnerships with workers, characterized by democratic procedures, active communications, and positive incentive systems. This technology is critical to ensuring that the manufacturing enterprise has the flexibility to change. The third area is hardware and software technologies that support skilled and knowledgeable workers (instead of replacing them) and that are usable in high-performance production environments. Turner noted that this last area is based on a recognition that workers are a source of creativity, not just an uncontrollable source of variation. He believes that an integrated perspective focused on people and systems would maximize peoples' capabilities, opportunities, and participation.

The research areas that would support this integrated perspective are listed below:

- the development of usable systems for representing and analyzing social systems
- the identification of barriers that stand between average and best practices
- the development and validation of usable tools for collaborative design, such as tools that include the workforce in the integrated process and product development (IPPD) process and enable cross-cultural systems of information management, representation, and communication
- the creation of a science of high-performance systems
- cross-disciplinary studies of the economics, behavioral, and social aspects of designing, implementing, and sustaining high-performance work systems
- the development of technologies that enhance workers' control over production, workplace organization, machinery, equipment, and technologies
- the development of metrics and methodologies for high-performance practices, participatory design, skill assessment, and worker-centered control
- the development of educational curricula

Turner ended his talk by predicting that the integrated approach he described would raise the standard of living, reduce social polarization, and strengthen democracy in the workplace and community. Technologies that focus on the people who operate the system will be critical to the success of manufacturing.

ORGANIZING MANUFACTURING WORK: AN INDUSTRY PERSPECTIVE

Steven J. Bomba

Johnson Controls, Glendale, Wisconsin

Steven J. Bomba presented an industry perspective on the organizational elements of manufacturing in 2020. He began by stating that the future does not just happen; it is made, and we, therefore, are responsible for how it is made. Manufacturing is an activity of the people, for the people, and by the people. In Bomba's view, people are not only the problem, but they also contribute to the solution.

Historically, people have changed slowly, requiring a generation to complete an adjustment. People can only change at their own pace. Unfortunately, they also become more expensive for businesses with time because of inefficiencies and medical costs, among other things. In the future, people will retain control of manufacturing processes and will have increasing expectations.

Like people, machines can also work, sell, and buy. Although they become less expensive with time and can be controlled, they are becoming increasingly complex as a result of new technologies. Machines are not yet self-organized. According to Bomba, the complexity of the human/machine relationship, and therefore of the manufacturing environment, is also increasing; the level of human/machine interdependence is increasing; and consequently, controllability by people is becoming more difficult, which increases risk.

Despite the increased risk, we have not been teaching management how to deal with risk. Instead we have been advancing incrementally. We need to diffuse a new technology into the marketplace more rapidly. Historical data prepared by Taniguchi (1983) showed that it took 30 to 50 years for metal-cutting technologies to diffuse from the laboratory to the factory (see Figure A-8). Technological discontinuities have led to quantum improvements in many industries. Although continuous improvement is necessary, it is not sufficient for survival. Organizations also need innovation. If organizations remain faithful to old traditions and don't take risks, they will surely fail.

Bomba believes that key factors for the success of manufacturing enterprises in the future will be effective team learning, accommodation of social differences, and effective leadership. His vision of a successful company is a small manufacturing unit that is customer-centered, cross-functional, and based on shared information and shared knowledge. But how will small manufacturers pay for research? A successful manufacturing system will have to be "genetically"

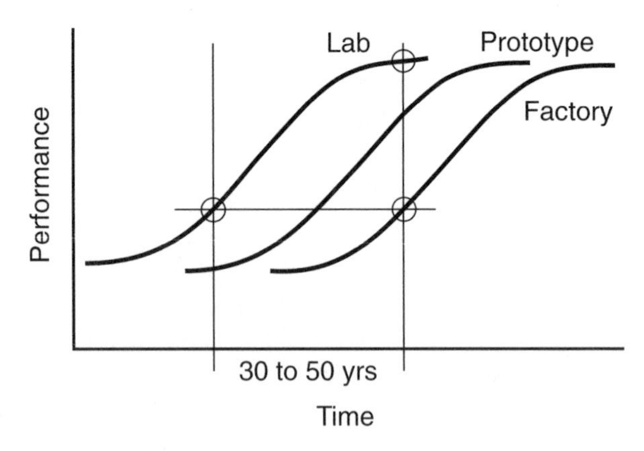

FIGURE A-8 The technology diffusion cycle. Source: Taniguchi, 1983.

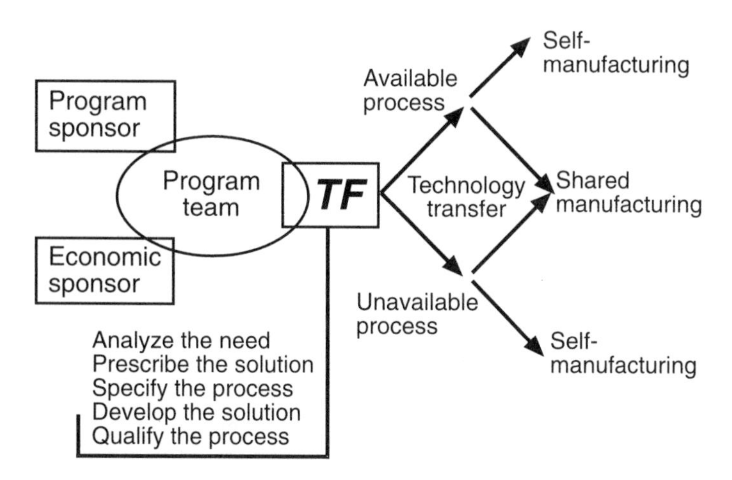

FIGURE A-9 The business organization of future enterprises.

diverse, with selective mutation, distributed strength, and "portfolioism" to enable small companies to bargain collectively.

Bomba emphasized the need for knowledge workers. Customizing products will require workers trained in an educational system based on an understanding of how people learn and geared toward individual learning. Workers must have access to just-in-time educational tools, as well as education based on past experience.

Organizations will have to be both reflexive and thoughtful, centralized and widely distributed. Tensions are sure to mount as the experience base clashes with new concepts and emerging social characteristics. Bomba warned of the potential pitfalls of taking just-in-time tools and outsourcing to extremes. Worker education must ultimately contribute to lower price, higher quality, and greater customer satisfaction.

Bomba's model of a business organization of the future consists of a program team sponsored by both a program advocate and an economic advocate (see Figure A-9). The objective of the program team is to transfer technology, either by self-manufacture with available or newly developed processes or by shared manufacture through partnerships. The ideal team will be based on shared information, shared knowledge, and innovation. Tomorrow's industrial engineer will build mediated teams, will produce "informated" workers, and will be a systems synthesizer. Bomba concluded by reminding the workshop that the 30-year-olds of today are the people who will build the future.

Reference

Taniguchi, N. 1983. Current status in, and future trends of, ultraprecision machining and ultrafine materials processing. CIRP Annals 32(2): 573–582.

KNOWLEDGE AND INNOVATION MANAGEMENT

Debra M. Amidon

Entovation International, Wilmington, Massachusetts

Debra Amidon began her presentation with examples of resistance to change. Reactions to the phrase, "you must change," which implies that workers do not have the necessary skills or education to continue doing their jobs, are usually resistance and fear. Reactions to the phrase, "you must innovate," which implies that workers' skills and accomplishments are still valid and that their proven competencies can be applied in new ways, are much more positive. Amidon then outlined the three main messages of her presentation: we are moving into a "knowledge economy" in which knowledge and innovation management will be essential; innovation can be managed; knowledge and innovation management will be part of the larger picture of manufacturing.

The knowledge economy will differ from its predecessors, the agricultural economy, the industrial economy, and the short-lived information age. Previous economies were focused on managing things outside of ourselves, whereas the knowledge economy is based on managing things inside ourselves, like the ability to create ideas and put them into action. The knowledge economy will require tapping into the intuition, intellect, and imagination of each and every participant.

Amidon presented evidence that the knowledge economy has already been embraced by the publication, *The Economist*, the Organization for Economic Co-operation and Development (OECD), and the National Research Council (NRC). According to *The Economist*, the knowledge economy is one of abundance. *The Economist*'s 1996 World Economic Survey states, "Economic theory has a problem with knowledge: it seems to defy the basic economic principle of scarcity . . . the more you use it and pass it on, the more it proliferates . . . [it is] infinitely expansible . . . What is scarce in the new economy is the ability to understand and use knowledge." According to Jean Claude Paye, the secretary-general of OECD, "The OECD is therefore devoting considerable effort to developing better indicators for knowledge inputs such as R&D and training expenditures, skills and competencies, flows of knowledge in the form of exchanges of ideas and diffusion of technology and, most of all, returns to knowledge investments."

The NRC's Productivity Paradox modeled the diffusion of knowledge and concluded that processes at the individual, group, and organizational levels greatly affect one another. Improvements in productivity at one level, for example, lead to much more than improvements in productivity at higher levels. Unfortunately, companies that are downsizing have generally ignored these organizational links. The people and institutions of the world are interconnected, and every nation has a vested interest in the productivity of other nations (NRC, 1994).

Amidon believes that the knowledge economy will necessitate a new (fifth generation) organizational focus (see Figure A-10). In the past, the organizational

		1st product as the asset	2nd product as the asset	3rd enterprise as the asset	4th customer as the asset	5th knowledge as the asset
	Core strategy	• Function association	• Link to business	• Technology/ business integration	• Integration with customer R and D	• Collaborative innovation system
	Change Factors	• Unpredictable serendipity	• Inter-dependence	• Systematic management	• Accelerated discontinuous global change	• Kaleidoscope dynamics
Management Operations	Performance	• Function as overhead	• Cost sharing	• Balancing risk/reward	• Productivity paradox	• Intellectual capacity impact
	Structure	• Hierarchies function as driver	• Matrix	• Distributed coordination	• Multidimensional communities or practice	• Symbiotic networks
	People	• We/the competition	• Proactive cooperation	• Structures collaborator	• Focus or values and capability	• Self managing knowledge workers
	Process	• Minimal communication	• Protect-to-protect basis	• Purposeful R and D portfolio	• Feedback goods and information persistence	• Cross-boundary learning and knowledge flow
	Technology	• Employers	• Data based	• Information based	• Has a competitive weapon	• Intelligent knowledge processors

Customer retention → Customer satisfaction → Customer success →

FIGURE A-10 Characteristics of business generations. Source: Rogers, 1996.

focus has been on products, projects, enterprises, and customers as the primary business assets. The new focus will be on knowledge as the primary asset. In the knowledge economy, organizations will be dealing with "kaleidoscope dynamics," i.e., rapidly changing and multifaceted business environments. Unfortunately, most organizations today still rely on outdated management technologies focused on projects and enterprises (Rogers, 1996). Business leaders in this new economy will have to be learners.

Knowledge management is not a fad, as evidenced by the considerable income being generated by knowledge management companies like McKinsey, the Big Six, and IBM. A "community of knowledge practice" is emerging. John Seely Brown, director of Xerox Corporation's Palo Alto Research Center, describes this community as "peers in the execution of 'real work' . . . What holds them together is a common sense of purpose and a real need to know what each other knows" (Brown and Gray, 1995). A detailed outline of the transformation of each business function (e.g., finance, human resources, quality, information technology, research and development, manufacturing) can be found in Amidon's recent book, *Innovation Strategy for the Knowledge Economy: The Ken Awakening* (Amidon, 1997). Amidon believes that "All are coming to a common language and shared purpose."

According to Peter Drucker, every organization, not just business organizations, needs one core competence, innovation, and the capacity to appraise innovative performance (Drucker, 1995). Charles Handy, author of *The Age of Paradox,* believes that companies must change while they are successful. Once they begin to have problems, he says, it is too late to change. A number of current business strategies, including concurrent engineering, agile manufacturing, and the strategy profession, do lead to innovation. The systematic management of innovation is based on the following interdependent factors:

- managed process collaboration
- systematic measurement of performance
- research and education for business development
- distributed learning network
- competitive intelligence
- value-added knowledge products/services
- strategic alliances and other innovative practices
- marketing campaign
- leadership leverage of intellectual competencies
- computer and communications technology

The managerial standards evolving for the twenty-first century will be consistent with these factors.

Innovation is a value system rather than a value chain. The strategic business network of the future will include all stakeholders—customers, customers' customers, suppliers, alliance partners, distributors, and, in some cases, competitors—and knowledge will come from everywhere in the system. Customers will be considered sources of knowledge, rather than passive recipients of goods or services. Customer innovation, or "innovating with the customer," will be the source of economic wealth. Companies that focus not only on customer retention and customer satisfaction but also on customer success will be successful (see Figure A-10). Success will depend on a company's understanding of the unarticulated needs of customers and identifying unserved markets.

Finally, Amidon briefly touched on her concept of the "The Ken Awakening," a unifying term that transcends business functions, industries, and geography. Ken is an international term for knowledge, understanding, and range of vision. By the year 2000, we will have had another World's Fair in Germany and a Worldwide Innovation Congress. Just as the United Nations was created to manage political instability and the World Bank was designed to manage the worldwide flow of financial assets, a new infrastructure will have to be created to manage the worldwide flow of ideas.

References

Amidon, D.M. 1997. Innovation Strategy for the Knowledge Economy: The Ken Awakening. Boston, Mass.: Butterworth-Heinemann.

Brown, J.S., and E.S. Gray. 1995. After re-engineering: The people are the company. Fast Company 1(1): 78–85.

Drucker, P.F. 1995. The information executives truly need. Harvard Business Review. 73(January/February): 54–62.

National Research Council. 1994. Organizational Linkages: Understanding the Productivity Paradox. Washington, D.C.: National Academy Press.

Rogers, D.A. 1996. Challenge of fifth generation R&D. Research Technology Management 39(July/August): 33–41.

KNOWLEDGE DELIVERY FOR THE TWENTY-FIRST CENTURY

Richard Altman

Communication Design, Tempe, Arizona

In his presentation, Richard Altman described his experiences using information technology in education, or "knowledge delivery." He believes that information technology will be the foundation for knowledge delivery in the twenty-first century.

Altman, who has directed the development of private networks for business and industry, gave an example of knowledge delivery for the educational community from his own experience. He and his co-workers started the project with an Apple computer. They built an operating system on top of the Apple operating system, but it was still too difficult for the educational community to use. Therefore, on top of that operating system, they built appropriate software for users of different ages. They set up an educational system with customized curricula for teachers and students of different skill levels. The curricula were delivered on a wide-area network called EduNet, and every school could buy a connection to EduNet.

Eventually, Altman and his colleagues discovered that the educational community wanted a broader array of products. At this point, ownership of educational materials became an issue. Altman's group went to the publishers (Viacom and Simon & Schuster) for permission to use their materials but discovered that these companies did not own the contents. Obtaining digital information to pass on to educational organizations, required building a series of satellite-delivered interactions corresponding with the electronic network.

Altman next described the global expansion of the network. Film crews were sent around the world on "curriculum journeys" to gather material, and an arrangement was established with the NOW channel for broadcasting them. Educators can now call and request material on almost any topic and they can preview the materials via satellite connections before making their selection. The next logical step, Altman said, is that textbooks will become obsolete.

The new learning tools incorporate multiple subject matter and multiple channels. The demand for these new technologies in business and industry will be focused on three areas: day-to-day communications, training, and formal education.

Altman noted that education in the future will be tailored to the individual. Every human being is capable of learning, he said, and technology should be used to educate everyone, even people with so-called "learning disabilities." Altman believes that with electronic educational tools we will be able to educate people who learn in unconventional ways.

As the technology evolves, education will also become more accessible at home. Companies can put a satellite dish in the home of every employee. One advantage of bringing educational programs into the home would be to reduce travel time. An example of effective home education cited by Altman was the continuing education courses for nurses in Florida. Nurses are required to take a certain number of credit hours each year to remain certified. In this program, professionals are available via satellite, and nurses can interact with them in person or electronically. The current problem is accreditation because universities have not been able to determine the number of credits for these courses. This type of education will, therefore, require others to rethink their ideas of education.

MANUFACTURING TECHNOLOGY AND ELECTRONICS MANUFACTURING IN 2010

Mauro Walker

Motorola, Schaumberg, Illinois

Mauro Walker made a number of predictions about manufacturing in general and the electronics industry in particular. He discussed the factories and workforce of tomorrow, the importance of information and technology development to the manufacturing enterprise, and changes in the organizational structure and level of globalization.

According to Walker, the factories of tomorrow will be highly flexible, productive, and capable of great variations in capacity. They will be structured to manufacture a large and varied mix of products, and they will be capable of rapidly realizing new products. Shortened product life cycles will demand high speed and the capacity for representative prototyping. The manufacturing line will be required to accommodate the introduction of new products with minimum modifications to manufacturing processes or equipment. Advancements in rapid manufacturing processes and tools will be introduced along with new products.

The manufacturing workforce of tomorrow will be characterized by high productivity and advanced skills, including knowledge of business and information systems. Direct human value-added in manufacturing will all but disappear because advanced tools will be responsible for increases in worker productivity. Internal training will focus on high-level professional education through alliances with universities worldwide.

Manufacturing information will be the new key resource, and the manufacturing enterprise will be capable of real-time worldwide communication on all levels, including production schedules, information about the supply chain, and plant capacities. Planning and scheduling will be immediate and accurate, and factories, production lines, and equipment will be simulated with complete accuracy. In addition, factory software control will provide most of the required flexibility on relatively standard mechanical platforms.

Walker predicted that there will be revolutionary advances in information, equipment, and materials technology and that most of these advances will result from collaborations on a number of levels of the technology supply chain and basic research. These collaborations will require the sharing of information and technology among competing manufacturers, thereby eliminating differentiations based on the development of manufacturing technology. Instead, differentiation will be based on first deployment and superior applications of technology. The key technologies for the electronics industry in the near future will be flexible and scaleable equipment and software, data-driven deposition techniques, high-density direct wire interconnects, product and factory modeling, and direct chip attachment.

The manufacturing organizations of tomorrow will be fewer in number, smaller, and virtual. Because of the breadth of knowledge of the highly-productive professional workforce, companies will need fewer and smaller internal organizations. Instead, virtual internal and external organizations will be based on specific products and functional objectives. Companies will, therefore, need systems that enable the functioning of virtual organizations. In addition, Walker believes that factories will be located around the globe, based on proximity to the market and availability of skilled workers. Factory locations will be flexible so companies can make the best use of fixed assets in a changing environment.

Finally, Walker spent some time describing the National Electronics Manufacturing Initiative (NEMI), an industry-wide project that was intended to complement SEMATECH, which focuses on material and equipment technologies. NEMI focuses on packaging, interconnects, and supply chain technologies and some auxiliary topics, such as storage, optics, and displays.

COMPLEXITY THEORY
NEW WAYS TO THINK ABOUT MANUFACTURING

Richard Morley

Morley and Associates, Milford, New Hampshire

Richard Morley introduced chaos theory by reminding workshop participants that people have always been too conservative in predicting the future; some of the brightest individuals of the past have predicted that certain inventions were impossible, only to be proven wrong. This is because people are trapped in the present, unable to see beyond their current paradigm. History shows, however, that progress is nonlinear, and it is impossible to make a smooth extrapolation into the future. Most people see the future as filled with chaotic situations, and therefore problems seem impossible to solve. Chaos abounds in nature, however. The movement of smoke, the flow of streams, and the creation and movement of weather patterns are all chaotic phenomena.

Some of the most chaotic human systems actually operate according to a few simple rules. For example, conventional wisdom has it that the dispatcher knows where the cabs are and that he directs available cabs where they are needed. This is not the case, however. In reality, taxis taking fares follow the basic "I'm empty, you need a ride" rule. This system is effective because most people can get a ride within a short period of time anywhere in the city. Another example is a dozen elevators servicing a 72-floor building. According to conventional wisdom, the computer dispatches the elevators and optimizes their use. In reality, these elevators operate very effectively according to the rule that the closest elevator answers a call and stops at that floor only if the call is for the same direction as the direction of the current passengers and only if there is room for more passengers.

According to Morley, there is no agreed-upon definition of complexity. However, he gave a list of systems that are generally considered to be complex, including DNA, the immune system, the brain, fluid turbulence, economies, and manufacturing. He described how nature develops systems for managing complexity by following a few simple rules that enable individual members to act together in a way that demonstrates collective intelligence. He cited the example of a flock of birds that can maneuver around buildings and trees without breaking up the flock. The birds follow a few simple rules: head for the nest, stay a fixed distance from other birds, fly at a constant speed, and slow down at corners. This "group intelligence" seems to solve the very complex problem of hundreds of components working toward the same goal without central control. Morley believes that this "nobody-is-in-charge" approach can be applied to the complex system of manufacturing.

Morley then described "spontaneous order" or "emergent behavior." When many independent elements following simple rules interact, they create a new system. This system is robust, deterministic, bound but not predictable, easily computable, understandable, easily changed, and adaptable. The system is probably more intelligent than the sum of its parts and behaves in complex ways. He described the system for painting truck bodies at General Motors as an example. For each truck every paint booth bids on the job. There was no central control, and no one knew which paint booth would do which job. The system allowed the components of the paint process (robots) to decide how to paint the trucks and which portions to paint when. This method resulted in efficient, high-quality painting and saved millions of dollars. Other examples cited by Morley included: a power plant boiler control system developed in a week with only 120 lines of code; the Baltimore Highway Control System developed in two weeks with 718 lines of code; and a self-managing assembly plant control system developed in five days with 632 lines of code. Each of these systems illustrates that complexity can be managed most efficiently by minimizing the number of rules. The fewer rules there are, the easier it is to create the overall control system and the easier it is to change it. The main obstacle to the widespread implementation of complex systems is resistance to changing the paradigm that complex systems require complex control systems to manage them.

GROUP DISCUSSIONS

Group assignments on the second day were different from day one and were as follows. The first name indicates the committee member who acted as facilitator for the group and the names in italics indicate the spokespersons who presented the results:

Group 1: Ann Majchrzak, Charles Carter, George Hazelrigg, M. Eugene Merchant, Mike McEvoy, Brian Turner, Patricia Whitman
Group 2: Barbara Fossum, William Hanson, Richard Jarman, *Richard Kegg*, Louis Kiefer, Rakesh Mahajan, Kathryn Whiting
Group 3: David Hagen, Debra Amidon, Rick Dove, *John Decaire*, David Miska, Leo Plonsky, Heinz Schmitt
Group 4: Eugene Wong, Nathan Cloud, Thomas Crumm, H.T. Goranson, Woody Noxon, Wilfried Sihn, *James Solberg*, Gordon Forward
Group 5: Donald Frey, Richard Altman, *Steven Bomba*, David Hardt, Robert Hocken, Richard Morley, Richard Neal
Group 6: Lawrence Rhoades, Bill Kay, Howard Kuhn, *Eric Larson*, Edward Leamer, F. Stan Settles, John Bollinger

The groups were asked to consider the following questions:

1. What are the top technical challenges to the achievement of our vision for manufacturing in 2020 (including enabling technologies and manufacturing technologies)?
2. What research and development should be done now to meet these technical challenges?

GROUP ONE

The group identified six categories of technological challenges:

- tools for simulation, planning, and design
- intelligent communication systems
- conversion processes and tools
- sustainability
- materials
- new products for which manufacturing processes still need to be developed

Some of the group members believed that another important technical challenge is the capacity to visualize organizations, interactions, and other complex processes. For example, before learning tools can be incorporated into the process, there must be an understanding of the process as a whole.

The following technical challenges and research and development areas were developed for each category:

Tools for Simulation, Planning, and Design

Technical challenges

- integration of design and manufacturing that includes detailed modeling of manufacturing processes to enable joint optimization of design, manufacturing, and organization
- incorporation of learning within tools
- incorporation of training within tools
- simulation of organizational issues
- development of theory/science of engineering design, manufacturing, and subcontractor management that includes values and preferences
- development of planning tools (e.g., simulating new business processes, market forecasting, and factory planning)
- incorporation of methods to accelerate the characterization of materials for production
- development of simulation-based learning tools (e.g., simulators and virtual reality) for current and future (K–12+) workers

Research and development

- standards for software compatibility or robust software that does not need standards
- transparent systems understandable to everyone
- methods to make data accessible to everyone (protocols, security, format, interoperability)
- information filtering
- representation of social and organizational processes across cultures in formats accessible to nonexperts
- intelligent agents
- interactive, 3-D, simulation-based visualizations of complex structures integrating behavioral, organizational, and people issues with other analyses
- using sound and color for pattern analysis
- methods to merge historical data with simulation systems
- simulation of alternative business processes
- methods to capture and catalogue development and problem-solving decision processes for real-time data retrieval

Intelligent Communication Systems

Technical challenges

- involving of all enterprise operations in information exchange
- systems compatibility between subcontractors and partners

- global systems to enable communication and cooperation superior to face-to-face methods
- incorporation of learning within communication systems
- incorporation of training within communications systems
- systems that deal with the nondeterministic nature of manufacturing
- incorporation of scheduling systems that allow for coordination of autonomous holonic agents (vs. centralized scheduling systems) with few simple rules (e.g., chaos theory)
- techniques that take advantage of the village cooperative (e.g., a small factory in the Philippines that assembles motors for elevators without advanced technologies)

Research and development

- large-scale, real-time simulation
- mind-to-mind communication (e.g., analysis of brainwaves, intentions)
- increased bandwidth/data compression
- information filtering (method for figuring out what needs to be filtered for different uses)
- delivery system and interfaces that can accommodate individual styles and preferences
- methods for remote transfer of skills
- remote access to experts (syndicated experts, centers of excellence)

Conversion Processes and Tools

Technical challenges

- processing tools for flexible and customized manufacturing
- process technologies to produce lot sizes of one competitively
- basic understanding of conversion processes to allow modeling and simulation
- incorporation of learning and real-time training
- equipment and methods for portable manufacturing
- sensors for process controls in closed-loop systems
- equipment and methods for small-scale manufacturing
- manufacturing processes that can "grow" products
- processes that can create products using ultrafine particles
- sensory feedback and data transmission technologies to enable remote manufacturing
- application of rapid prototyping technology for designing and producing tooling

Research and development

- integrated systems that combine software, sensors, and actuators
- process models and supporting data for conversion processes
- intelligent process representations and models for conversion processes
- intelligent process algorithms that can create flexible process models
- smarter equipment that uses modeling and simulation tools for learning
- remote sensing of human feedback
- faster nanotechnology processes

Sustainability

Technical challenges

- new technologies for handling manufacturing process waste
- methods of portable manufacturing for reclaiming process waste
- incorporating environmental sustainability into engineering design processes
- manufacturing practices and policies that support sustainable, global environments
- heat exchangers and efficient co-generation processes to recovery energy from waste
- more energy-efficient products and systems

Research and development

- lighter, smaller equipment
- efficient manufacturing processes with reduced scales of operation
- simulations and databases for engineering design
- methods and data that can predict the effects of alternative manufacturing processes on the global environment

Materials

Technical challenges

- advanced nanoparticle materials
- applications of genetic engineering to high-volume manufacturing
- materials that decompose to elementary particles

New Products

Technical challenges

- portable energy storage (e.g., fuel cells and polymer batteries)
- human interface components (e.g., speech recognition)
- mass memory storage technology (e.g., giant magneto-resistance)

GROUP TWO

Question 3: Technical Challenges

Produce and maintain a workforce with the training and capability to add high value. To gain maximum value from the workforce, manufacturing enterprises must integrate human capital into their business processes; acquire and use knowledge more efficiently and effectively; make the innovation process more efficient and effective; create the pull for knowledge; reward and create incentives for learning; develop affordable group education tools that can change behavior (e.g., interactive computer learning); determine how to make prescriptive learning more effective and affordable; and determine the costs and benefits of knowledge acquisition and learning.

Foster innovation. Rapid change could cause significant problems for manufacturing. Manufacturing enterprises will have to foster innovation among all employees; create an environment that encourages innovation; link innovation to business strategy; and teach and apply creative thinking skills.

Provide real, physical experiences to supplement simulations for training. Some participants were concerned that simulations would not provide realistic experiences for teaching operators to run processes. Physical experiences should also be integrated into learning environments.

Resolve problems of connectivity and data representation. This will require expediting digital design data to the shop floor (for process planning and control); providing easy-to-use and easy-to-connect computer systems; and creating flexible automation that can receive and use digital design data.

Question 4: Research and Development

After identifying the technical challenges, the group discussions focused on research issues. The following research opportunities were identified by members of the discussion group:

- greater bandwidth for communication systems
- the equivalent of generally accepted accounting principles for human capital, including quantifying human knowledge; quantifying the value of knowledge alliances, partners, suppliers, and customers; and calculating the economic value of industrial training
- methods and tools to facilitate decision-making processes, including consensus decision making, managing risk, and managing collaborative projects (i.e., "alliance tools")
- multimedia electronic learning based on the most current knowledge about the learning process; shells for subject-matter experts to develop techniques and tools
- methods to expedite digital design data to the factory floor (e.g., for pro-

cess planning and production), including software and hardware that is connectable automatically and can use digital design data without human intervention
- integrating creative thinking skills into management practices

GROUP THREE

Question 3: Technical Challenges

Group discussions initially focused on identifying the top technical challenges for manufacturing. A significant amount of time was spent discussing the interplay and interdependencies of the following three technical challenges:

- creating, designing, and exploiting knowledge systems
- developing real-time, on-demand learning at individual, team, and company levels and tailoring course designs and delivery methods to the learning modes of the students
- developing information technologies, including network and user interfaces, software libraries for manufacturing, "plug and play" systems, and software productivity

Other challenges that received strong support from individual participants included the following:

- revolutionizing unit process technology with quantum leaps in process capabilities.
- the capability to predict product reliability
- assessing collaboration strategies and processes to determine the best ways to develop and implement collaborative relationships
- designing and managing reconfigurable factories

Additional challenges that were suggested by individual participants included the following:

- managing nonlinear systems
- measuring performance holistically
- developing life cycle engineering approaches to design reusable, recyclable products economically
- integrating product and process designs

Question 4: Research and Development

The research and development areas discussed are listed below:

- techniques to convert tacit knowledge to explicit knowledge that is usable at several levels (e.g., individual, team, and organization levels)

- determination and identification of individual learning styles to facilitate the development of appropriate learning materials and delivery systems with emphasis on real-time and on-demand learning
- system-independent knowledge representations that can distribute knowledge throughout the entire manufacturing enterprise
- breakthrough technologies in free-form fabrication, micro-manufacturing, nanomanufacturing, and biomanufacturing processes
- simulation technologies that can predict product/process reliability
- tools for collaboration

GROUP FOUR

Various members of the group discussed the following technical challenges for manufacturing in the year 2020:

- implementing computer-based information systems for modeling, synthesis, optimization, and on-line control of manufacturing, from the process level to the enterprise level
- representing human components, not only for accurate modeling, but also for feedback on an individual's impact on the overall system
- developing mechanisms of self-organization for manufacturing organizations in a variety of settings (e.g., self-assembling teams of workers with limited skills that exhibit a high degree of collective capability for efficient manufacturing and solving the social problem of the have-nots)
- developing robust design methodologies that can accommodate technological and market changes
- developing a science-based understanding of the physical phenomena (mechanical, thermal, and chemical) that occur in manufacturing unit processes
- considering alternative manufacturing paradigms (e.g., non-assembly-line approaches and customer-performed manufacturing)
- developing a "language" that describes manufacturing in terms of basic production operations and rules (syntax) that can represent the manufacturing process as a program and represent actual manufacturing as the execution of the program (A very high degree of flexibility is achieved in this way because the same parameters and syntax can describe a wide variety of products.)
- developing biomanufacturing processes based on genetic engineering techniques
- applying life cycle engineering approaches to the development of environmentally-friendly manufacturing that considers the entire life cycle of a product, including disposal

GROUP FIVE

Question 3: Technical Challenges

Discussion participants suggested that the technical challenges for manufacturing in 2020 will be to manufacture products "cheaper, better, and faster." The most important characteristics of manufacturing will be precision, speed, control, and cost. Precision and speed can increase value; control and cost can decrease costs.

Question 4: Research and Development

The group participants identified the following areas for research and development:

- real-time enterprise controls and interoperability standards and protocols for complex systems
- process controls based on parallel (computer) processing (fractal systems design) rather than sequential processing (von Neumann)
- microscale processes and machines (e.g., data links and sensors) including focused, extreme-UV precision processes, molecular assembly, and atomic processes
- biomanufacturing processes for the food, drug, and chemical industries (Agro-based chemicals [biomanufacturing/processing] are already being produced [e.g., growing insulin in alfalfa]. Medical implants were discussed, including "add-plants" or implants that will grow after implantation.)
- processing technologies for personal, neighborhood, and point-of-sale manufacturing (The group considered these concepts to represent the shifting economies of scale.)

GROUP SIX

The discussion participants identified a number of technical challenges and related research and development areas to realize the goals of visionary manufacturing for 2020: engineering the "socio-technical interface"; finding and keeping high-performance workers; constructing high performance work group and organizational/enterprise structures; providing materials/process/product modeling at all enterprise levels; optimizing the use of information/knowledge; reducing the "footprint" of manufacturing processes; and determining the roles of local government and business in education.

Engineering sociotechnical interfaces. Many discussion participants felt that it would be important to understand the role of the sociotechnical interface (i.e., the soft factors that can enable more effective and efficient manufacturing processes) at the individual, group, and enterprise levels. Some participants suggested that research and development should begin with reliable computer simulation models that systematically relate the value of soft factors to the product. Unless these soft factors are accounted for in the bottom line, they are likely to be undervalued by manufacturers. Soft factors include: the quality of human capital (e.g., education, skill sets, and intelligence); education and training programs to improve worker productivity and performance; and other human and group factors that contribute to high-performance organizations. Interdisciplinary teams (e.g., engineers, industrial psychologists, and economists) should develop models relating soft factors to costs.

Finding and keeping high-performance workers. Visionary manufacturing enterprises will be competing for, hiring, training, and refreshing the skills of the most-qualified employees. Manufacturing enterprises will also be concerned with protecting and quantifying the value of knowledge imparted to workers through education and training. As a consequence, it will become increasingly important to understand how employees learn so that knowledge can be developed, maintained, and refreshed cost-effectively.

Constructing high-performance work groups. Visionary manufacturing enterprises will have to combine highly skilled individuals from different cultures and with different educational backgrounds, skills, personalities, and styles in ways that will foster highly productive work groups.

Creating high-performance organizations/enterprises. At the enterprise level, visionary manufacturing enterprises will constantly strive to optimize the balance of manufacturing technologies with human/group factors to meet performance goals. Manufacturing technologies will have to be adaptable to evolving organizational structures, product lines, and processes. Enterprises will require near real-time measurements of outcomes, including a far wider range of measures than are currently used. Finally, firms will need incentive structures (e.g., equity arrangements, performance-related bonuses) to ensure that employees have a strong stake in the performance of the enterprise and to protect the knowledge and skills valued by the enterprise.

Providing materials/process/product modeling at all enterprise levels. Simulation models will have to link materials, processes, and products at all levels: molecular, discrete, and continuous. Visionary firms will use these models and systems to integrate product designs, materials, and process life cycles. Models could also include social and economic considerations that can identify the best candidates for jobs and combine individuals to form optimal work groups.

Optimizing the use of information/knowledge. Information and knowledge will be important to future manufacturing enterprises. For example, enterprises will have to understand fundamental scientific principles, readily available materials/processes/product information, materials properties, and other manufacturing information. Research should focus on the analysis, synthesis, and problem-solving capabilities of enterprises, how these relate to organizational culture, processes, and tools, and how these capabilities can be nurtured.

Reducing the "footprint" for manufacturing processes. Manufacturing technologies in the future should be small, inexpensive, adaptive, highly flexible, and redeployable. The goal is to improve efficiency and ease of use and to reduce power consumption.

Determining the roles of local government and business in education. Many of the discussion participants perceived a widening gap between the growing need for highly skilled job candidates and the apparently diminishing ability of the public education system to produce these candidates. This led to a discussion of public and private roles in education and the responsibilities of educating and training the future workforce and the fundamental issue of who should pay for kindergarten through "*n*th" grade (the group was uncertain what the value of *n* should be), and for technical and scientific education in secondary and higher education. Some participants suggested hybrid options, such as partnerships between industry and school districts, that might ensure the availability of individuals with the education and skills necessary for manufacturing jobs. Some participants also noted that because educational performance was closely related to family (especially parental involvement) and socioeconomic circumstances, employers could develop incentive systems to nurture better parenting, teaching, and academic performance.

APPENDIX B

Delphi Survey: Methodology and Results

INTRODUCTION

As part of its data gathering effort, the National Research Council (NRC) Committee on Visionary Manufacturing Challenges developed and implemented a survey. Using the Delphi method, international experts in manufacturing were surveyed to obtain a forecast of future manufacturing challenges for the year 2020. The Delphi survey was undertaken during a six month period from February to July 1997.

THE DELPHI METHOD

The term "Delphi method" refers to a variety of group communication processes for forecasting or decision making. The basic concept originated in the 1950s at the Rand Corporation as a spinoff of Air Force-sponsored research on the use of expert opinion. The original study involved a series of questionnaires with controlled feedback to determine the opinion of a group of experts on the U.S. industrial systems most likely to be targeted by Soviet strategic planners. At that time, the alternative would have been an extensive and costly data-collection process that included programming and executing computer models that could not be handled by available computers (Linstone and Turoff, 1975). During the past three decades, the Delphi method has been used by corporations, universities, government agencies, and nonprofit organizations for planning, technical and strategic evaluations, and forecasting.

The Delphi method characteristically obtains independent inputs from a group of individuals through an anonymous, iterative survey with controlled feedback after each iteration. Delphi participants may be experts or laypersons depending on the goals of the survey. In most cases, the first questionnaire poses the problem in broad terms and invites answers and comments. Responses to the first questionnaire are then summarized and used to construct the second questionnaire, which presents the results of the first and gives participants an opportunity to refine their responses, clarify issues, identify areas of agreement or disagreement, and develop priorities. This interactive process can be repeated as many times as appropriate (Ziglio, 1996).

The Delphi method is widely considered to be effective in situations where no hard data exist and the primary source of information is well informed, learned opinion (A.T. Kearney, Inc., 1988). Experiments carried out in the late 1960s and early 1970s demonstrated that the Delphi method has distinct advantages over traditional, interactive group processes when the best available information is the judgment of knowledgeable individuals, (Dalkey, 1969; Ziglio, 1996).

SURVEY ON VISIONARY MANUFACTURING CHALLENGES

The NRC Delphi survey on visionary manufacturing challenges was designed to forecast manufacturing challenges for 2020 and to identify enabling technologies for research and development. The work of implementing the survey included designing and testing the first questionnaire; identifying, selecting, and contacting potential participants; distributing the first questionnaire; collecting and analyzing responses from the first questionnaire; designing the second questionnaire; distributing the second questionnaire; and collecting and analyzing the responses from the second questionnaire.

Designing and Testing the First Questionnaire

In February 1996, a Workshop on Methods for Predicting Manufacturing Challenges was held at the Beckman Center in Irvine, California. The workshop was conducted by the BMAED as a means of determining the best methods of gathering data for the study on visionary manufacturing challenges. At the workshop, participants from the United States, Europe, and Japan took part in a roundtable discussion and filled out a trial questionnaire. The results of this workshop and recent questionnaires on manufacturing issues were used to prepare the first questionnaire of the BMAED Delphi survey.

The purpose of the first questionnaire was to elicit information on participants' visions of (1) the competitive environment in 2020, (2) characteristics of manufacturing enterprises in 2020, (3) the challenges that would be faced by manufacturing enterprises, and (4) the technological developments that would enable manufacturers to meet the challenges. The committee decided to use questions calling for open-ended responses, as opposed to providing respondents with a selection of answers to choose from. This was done to encourage creative thinking on the part of respondents and to ensure that the scope of survey responses was not limited to the committee's knowledge and thinking. A copy of the first questionnaire is provided in Appendix C.

The first questionnaire was pilot tested on seven individuals identified by committee members as having suitable manufacturing experience, vision, and familiarity with the project, as well as the ability to complete and return the pilot questionnaires quickly. The results of the pilot questionnaires were incorporated into the instructions and questions in the first questionnaire.

Selection and Composition of Survey Participants

Potential survey participants were identified using numerous mechanisms. Members of the Committee on Visionary Manufacturing Challenges identified both potential participants and individuals who could suggest potential participants. Members of BMAED and members of Section 8 of the National Academy of Engineering (Industrial, Manufacturing, and Operational Engineering) were contacted and asked to participate. Recommendations were also requested from national and international manufacturing organizations, including the Agility Forum, ASM International, the Consortium for Advanced Manufacturing International (CAM-I), the Coalition for Intelligent Manufacturing (CIMS), the Council on Competitiveness, the Fraunhofer Society of Germany, the Industrial Research Institute, the Institute of Electrical and Electronics Engineers (IEEE), the International Institute for Production Engineering Research (CIRP), Intelligent Manufacturing Systems (IMS), the National Association of Manufacturers (NAM), the National Center for Advanced Technologies (NCAT), the National Center for Manufacturing Science (NCMS), Next Generation Manufacturing Systems (NGM), the Society of Automotive Engineers (SAE), and the Society of Manufacturing Engineers (SME). Recommendations were also requested from the National Science Foundation (NSF). Table B-1 shows the number of participants from each source.

The criteria for selecting participants included manufacturing expertise and evidence of visionary thinking. Because of time constraints and to facilitate the analysis, the survey was conducted by facsimile and email. The list of potential participants was, therefore, narrowed to those who had either a working facsimile number or a working email address. The committee believed that the survey should include participants from both industry and academia as well as U.S. and international experts in manufacturing. Special efforts were made to contact a large number of international and industry participants.

As shown in Table B-2, the largest representation was from U.S. industry, followed by international and U.S. academia. The international academics were primarily located in Europe. The percentage of respondents from international industry was less than 10 percent. This was probably attributable to the composition of the original lists, which were focused on U.S. industry and international academics, and the difficulty in identifying representatives of international indus-

TABLE B-1 Number of Potential Survey Participants Identified and Contacted and Number Who Responded to Questionnaires 1 and 2

Organization or Method of Identification	Number of Potential Participants Identified [a]	Total Number Successfully Contacted for Questionnaire 1 [b]	Total Number of Respondents to Questionnaire 1	Percentage that Responded to Questionnaire 1	Number of Respondents to Questionnaire 2	Percentage that Responded to Questionnaire 2	Percentage of Questionnaire 1 Respondents that Responded to Questionnaire 2
ASM[c]	28	28	12	43%	10	36%	83%
BMAED	18	18	7	39%	7	39%	100%
CIRP	74	44	22	50%	20	45%	91%
IMS[d]	56	54	15	28%	12	22%	80%
NAE[e]	80	58	9	16%	9	16%	100%
SME[f]	468	149	43	29%	35	23%	81%
Individual Referrals[g]	258	203	73	36%	62	31%	85%
Total	982	563	181	32%	155	28%	86%

[a]If an individual appeared on more than one list, he or she was assigned to the list where the name first appeared.
[b]Some potential participants were not contacted because of incorrect or insufficient contact information.
[c]Members of ASM International were recommended by the president of that organization.
[d]Individuals in the leadership of Intelligent Manufacturing Systems (IMS) were selected.
[e]Members of the National Academy of Engineering's (NAE) Section 8 (Industrial, Manufacturing, and Operational Engineering) was contacted.
[f]Fellows of the Society of Manufacturing Engineers (SME) and members of the SME Boards of Advisors were contacted.
[g]Individual referrals came from committee members, the Agility Forum, the Coalition for Intelligent Manufacturing Systems (CIMS), the Consortium for Advanced Manufacturing - International (CAM-I), the Fraunhofer Society, the National Science Foundation (NSF), the National Association of Manufacturers (NAM), Next Generation Manufacturing Systems (NGM), and the Society of Automotive Engineers (SAE).

TABLE B-2 Distribution of Respondents to Questionnaire 1 by Country and Work Affiliation

	Respondents					
	Industry		Academia		Other[a]	
	Number	Percent	Number	Percent	Number	Percent
United States	63	35%	42	23%	4	2%
International	16	9%	51	28%	5	3%
Africa and the Middle East[b]	0	—	4	2%	0	—
Asia[c]	8	4%	13	7%	0	—
Australia	1	—	2	1%	1	—
Europe[d]	3	2%	29	16%	0	—
North America (non-U.S.)[e]	2	1%	3	2%	3	2%
South America[f]	2	1%	0	—	1	—

[a]This category includes government agencies and trade organizations.
[b]Countries represented were Israel and South Africa.
[c]Countries represented were China, Japan, and Singapore.
[d]Countries represented were Belgium, Denmark, France, Italy, Germany, Norway, Romania, Sweden, Switzerland, and the United Kingdom.
[e]Countries represented were Canada and Mexico.
[f]Country represented was Chile.

try. The committee attempted to elicit a significant contribution from China and South America, two economies that are likely to become increasingly important in manufacturing as 2020 approaches. Efforts were made to identify individuals from China, Brazil, and Chile. Unfortunately, only a few potential participants were identified, and the response rate from them was low. Reasons included the difficulty of contacting these individuals via facsimile or email, problems with language (the questionnaires were not translated), and the cost of responding via facsimile or email.

Implementation and Analysis of the Questionnaire 1

A letter describing the project was sent to potential participants prior to the first questionnaire. The purpose of the letter was to familiarize potential participants with the goals of the project and to send a personal request for cooperation from the committee chair. Approximately one week later, the first questionnaires were sent out.

The first questionnaire was sent out in batches via facsimile and email between February 15, 1997, and March 17, 1997, and participants were given approximately two weeks to respond. If time permitted, busy facsimile lines were retried and returned emails were resent. One reminder notice was sent to those who did not reply by the original deadline. Responses were received between February 21, 1997, and May 4, 1997. The final number of respondents, including the seven pilot respondents, was 181, or 32 percent of the individuals who were contacted (see Table B-1).

The responses to the first questionnaire varied in length and detail because of the open-ended nature of the questions. The technique of "open-coding," developed by Glaser and Strauss (1967), was used to analyze the responses. Using this technique, survey responses were read and reread, and codes, or categories, were inferred from them. Text sections containing similar phrases were grouped according to these codes. The idea behind the method is that the codes are not simply deduced from the analyst's ideas but are inferred from the survey responses.

Because of the time-intensive nature of this technique, the committee selected a consultant, Dr. Brian Borys of the School of Public Administration at the University of Southern California, to undertake the analysis. Several measures were taken to ensure that Dr. Borys' coding could be replicated and that the codes where consistent with the study objectives. First, Dr. Borys held preliminary discussions with committee members regarding the nature of the survey, the characteristics of the respondents, and the questions that needed to be answered. Second, Dr. Borys and several committee members separately coded survey responses drawn at random and compared their results. Consistency among the coders was sufficient to conclude that other coders could generate similar interpretations and that Dr. Borys' interpretive scheme would provide sufficient information. As a final check, the committee reviewed Dr. Borys' results after he had coded approximately half of the surveys and before he proceeded to code the rest.

When the coding was complete, the committee used the codes, or categories, to distill a list of manufacturing challenges and enabling technologies for 2020 that represented the ideas of the respondents to the first questionnaire. This list was then incorporated into the second questionnaire.

Design and Implementation of the Second Questionnaire

The Delphi method is an interactive process, i.e., during the process, participants receive feedback on the responses of the group as a whole. In the BMAED Delphi survey, the second questionnaire was used to provide participants with feedback on the results of the first questionnaire. The lists of manufacturing challenges and enabling technologies generated by the first questionnaire were used to construct the first two questions of the second questionnaire, which asked respondents to indicate the challenges and technologies they considered most important. Two additional questions were added asking respondents to list research topics based on their prioritized enabling technologies and the manufacturing challenges that would be addressed by these technologies. A copy of the second questionnaire is attached in Appendix C.

The second questionnaire was distributed via facsimile and email between May 16, 1997, and June 30, 1997, to respondents to the first questionnaire. The second questionnaire was also distributed to several individuals who had been unable to complete the first questionnaire but had expressed an interest in participating in the second. This questionnaire was also distributed to potential participants from China and South America who had not responded to the first questionnaire in an effort to increase representation from these two regions. Participants were asked to return the questionnaire within two weeks, and two reminders were sent. Responses were received between May 19, 1997, and July 14, 1997. Of the original 181 respondents, 155 (86 percent) returned the second questionnaire. In addition, nine individuals filled out the second questionnaire only.

Results from the Second Questionnaire

The responses from the second questionnaire were collated to determine the manufacturing challenges and enabling technologies that the respondents considered most important. These results are shown in Tables B-3 and B-4. Table B-5 shows how the respondents correlated enabling technologies and manufacturing challenges.

TABLE B-3 Manufacturing Challenges Prioritized by International Experts in Manufacturing

Identifier	Manufacturing Challenge	Votes	Rank
a	Enhancement of workforce performance and satisfaction to address rapidly changing and complex operational requirements and diverse culture-based issues	86	2
b	Constant, concurrent development of innovative products, processes, and systems to meet shorter product life cycles, enhance value added, and advance manufacturing capabilities	92	1
c	Ability to develop and execute complex and dynamic alliances and collaborations rapidly	52	5
d	Response to severe constraints on environmental impact and the increasing scarcity of materials and energy	66	3
e	Achievement of the speed and flexibility for cost-effective fulfillment of customer demands for instant satisfaction with customized products	45	6
f	Adoption of rapidly developing technologies to increase and/or adapt the core strength of the enterprise to the marketplace	40	7
g	Development of an effective global infrastructure to support optimal-scale manufacturing configurations	36	8
h	Effective conversion of information to useful knowledge in an environment where the volume of available information is increasing rapidly	64	4

TABLE B-4 Enabling Technologies Prioritized by International Experts in Manufacturing

Identifier	Enabling Technologies	Votes[a]	Rank[b]
A	Adaptable and reconfigurable manufacturing processes and systems (e.g., intelligent, mass customization; rapid creation of new production facilities; ability to accommodate a wide range of product characteristics)	86	1
B	Systems model for all manufacturing operations (e.g., real-time synthesis of planning; market demand; product development; distribution; social systems; wealth creation into manufacturing system planning; effective modeling of supply chains)	60	2
C	Micro- and nanotechnology for fabrication processes (e.g., atom-by-atom fabrication of assemblies; development of microscale machines)	32	9
D	Processes to customize totally new materials with order of magnitude property improvements designed on the atomic scale (e.g., an order of magnitude improvement in strength; defect-free materials; smart materials that can change properties in service in response to changing conditions; materials designed to be reprocessed or reconstructed)	45	7
E	Direct machine/user interfaces that enhance human performance and promote intelligent input (e.g., skill-leveraging, human commands transmitted directly to machine; human access to data via "bionic ears")	49	5
F	Net shape, programmable, flexible forming processes that require no hard tooling (e.g., pulsed power autoshaping; forming finished assemblies from the melt)	30	14
G	Design methods and manufacturing processes for products that can easily be reconfigured with software or hardware (e.g., products that are easily upgradable in the field for long life; products that can be customized by the customer)	51	4
H	Desk-top manufacturing processes (e.g., manufacturing in the home by customer; neighborhood manufacturing service centers; highly distributed manufacturing capacity according to market location; portable manufacturing)	20	18
I	Biotechnology processes for manufacturing (e.g., use of biological structures in engineering design; fabrication of parts and assemblies with biological processes; "designer" proteins, enzymes, and tissues; biocatalysts; bioassembly of new foods; biodevices for computer memories)	33	8
J	Scientific bases for manufacturing processes (e.g., rapid development of models for simulation)	32	9
K	Application of chaos theory to manufacturing (e.g., software that captures emergent behavior; developing basic rules of behavior in manufacturing systems; embedded intelligence software; negotiating and bargaining algorithms)	14	23

TABLE B-4 *Continued*

Identifier	Enabling Technologies	Votes[a]	Rank[b]
L	Waste-free manufacturing (e.g., processes designed with no by-products in manufacture, integrated multiple product lines that consume by-products of one line in another)	52	3
M	New transportation concepts for rapid movement of materials and products (e.g., friction reduction; antigravity; superfast conveyance)	9	24
N	Synthesis and architecture technologies for converting information into desired knowledge (e.g., human memory relational structures; capturing, synthesizing, relating, integrating, and systematizing new knowledge into applications-oriented uses)	47	6
O	Design methodologies that process a broader range (by orders of magnitude) of product requirements (e.g., include life-cycle design; producibility; societal requirements; workforce needs)	32	9
P	Unified methods of communication and protocols for the exchange of manufacturing enterprise information	31	13
Q	Processes for rapid and cost-effective development, transfer, and utilization of technology (e.g., innovation processes; new paradigms for technology development; analysis and synthesis of new technologies)	29	15
R	Methodology for quantum jumps in product and process reliability (e.g., variability reduction; new methods for robust design)	14	22
S	Low energy consumption processes (e.g., low-inertia machines; catalyst; alternate energy sources; high energy-density batteries)	29	15
T	New sensor technology for precision process control (real-time sensors for machine self-calibration; self-verification; self-correction; self-improvement)	32	9
U	360-degree collaboration software (e.g., translate neural knowledge base to language that is personalized to different thinking styles; enable workforce participation in technology design and development; interactive visualization)	16	21
V	Low gravity, high vacuum manufacturing (e.g., practical manufacturing in space; earth-bound manufacturing facilities with space environment)	7	25
W	New educational methods (e.g., in-home facilities; smart and knowledge pills)	21	17
X	New concept manufacturing processes (e.g., ion beam, three dimensional chemical etching)	18	20
Y	New software design methods (e.g., methods that are robust, seamless, adaptive, inter-operable, and highly reliable)	19	19

[a]Respondents could select more than one enabling technology to prioritize.
[b]In case of a tie, items were given the same rank.

TABLE B-5 Enabling Technologies for Meeting Manufacturing Challenges (Number of Respondents)

	Enabling Technologies																								
Challenge	A	B	C	D	E	F	G	H	I	J	K	L	M	N	O	P	Q	R	S	T	U	V	W	X	Y
a	27	20	2	4	31	0	13	10	1	4	6	8	1	19	17	12	8	3	0	7	7	0	10	2	7
b	51	34	12	21	12	17	26	10	11	15	4	8	0	15	20	12	13	6	3	17	6	2	3	10	14
c	20	22	1	0	6	0	9	5	3	5	6	37	0	14	8	13	5	1	2	0	11	0	3	1	6
d	14	11	8	21	2	7	12	6	18	11	3	3	0	8	11	4	8	1	17	5	4	2	2	3	3
e	50	26	2	14	11	13	21	8	7	9	4	5	1	14	18	13	10	3	2	14	8	0	2	6	9
f	30	16	4	6	4	11	15	2	7	8	3	3	2	10	12	6	15	1	1	8	6	1	3	5	9
g	11	20	1	1	4	0	7	9	1	4	2	6	1	10	7	15	4	1	1	0	7	1	1	0	4
h	16	16	4	2	15	3	12	3	3	6	10	9	1	29	13	19	10	2	3	6	8	0	7	0	11
Total	219	165	34	69	85	51	115	53	51	62	38	79	6	119	106	94	73	18	29	57	57	6	31	27	63
Rank	1	2	9	10	7	16	4	15	16	12	18	8	24	3	5	6	9	23	21	13	13	24	20	22	11
Number	8	8	8	7	8	6	8	8	8	8	8	8	5	8	8	8	8	8	7	6	8	5	8	6	8

REFERENCES

A.T. Kearney, Inc. 1988. Countdown to the Future: The Manufacturing Engineer in the 21st Century. Dearborn, Mich.: Society of Manufacturing Engineers.

Dalkey, N.C. 1969. The Delphi Method: An Experimental Study of Group Opinion. Memorandum RM-5888 PR. Santa Monica, Calif.: Rand Corporation.

Glaser, B.G., and A.L. Strauss. 1967. The Discovery of Grounded Theory. Chicago, Ill.: Aldine.

Linstone, H.A., and M. Turoff (eds). 1975. The Delphi Method: Techniques and Applications. Reading, Mass.: Addison-Wesley.

Ziglio, E. 1996. The Delphi Method and its Contribution to Decision-Making. Pp. 3–33 in Gazing into the Oracle: The Delphi Method and Its Application to Social Policy and Public Health. M. Adler and E. Ziglio, eds. London: Jessica Kingsley Publishers.

APPENDIX C

Delphi Survey Questionnaires

NATIONAL RESEARCH COUNCIL
COMMISSION ON ENGINEERING AND TECHNICAL SYSTEMS

BOARD ON MANUFACTURING AND ENGINEERING DESIGN
COMMITTEE ON VISIONARY MANUFACTURING CHALLENGES

Office Location:
Harris Building, Room 262
2001 Wisconsin Avenue, N.W.
Phone: (202) 334-3505
FAX: (202) 334-3718
Internet: bmaed@nas.edu
Mailing Address:
2101 Constitution Avenue, NW
Washington, DC 20418

SURVEY
MANUFACTURING GRAND CHALLENGES IN THE YEAR 2020

Dear Participant:

Thank you for contributing to this Delphi survey. Its purpose is to define the major challenges facing the manufacturing enterprise in the year 2020 and beyond and those technologies required to meet the challenges. This study should leapfrog current studies (such as that of NGM) and should not extrapolate from current trends. Results of this effort will help the National Research Council of the United States to identify technology research needs focused on areas likely to have significant impact on future manufacturing requirements beyond the year 2020. Results of this global survey can be obtained by any person in any country from the National Academy Press of the United States.

You were recommended for participation as a member of the international manufacturing community with the knowledge, vision, and insight needed to provide the original ideas required for this study. The survey has participants from Asia, Europe, North America, and South America. It provides you the opportunity to interact with this global community in defining technological directions with the potential for profoundly influencing manufacturing.

The Delphi survey process treats all inputs anonymously and uses iterations based on prior input to provide a focus for the results and to reach consensus. This survey will have three rounds of questions and will be conducted over a period of several months. The questionnaire for the first round is attached. It is relatively short with only four questions. Your value added will be reflected in the thoughtfulness of your answers.

Please return the completed questionnaire by February 28, 1997 so that it can be analyzed and incorporated in round 2. Thanks again for your contribution.

John G. Bollinger
Chair, Committee on Visionary Manufacturing Challenges

Robert M. Rusnak
Study Director

Respondent Code: ______
(To be filled out by NRC)

DELPHI SURVEY:
MANUFACTURING GRAND CHALLENGES IN THE YEAR 2020

QUESTIONNAIRE #1

Participant Information

Name:
Position:
Affiliation:
Products Manufactured: By your Division By your Company If your location is a research facility indicate the type of research performed
Approximate Size (number of employees): Of your Division Of your Company
Mailing Address: Industries that you are representing:
Telephone:
FAX:
e-mail:

Return by February 28, 1997 to:

NATIONAL RESEARCH COUNCIL
Board on Manufacturing and Engineering Design
Harris Building 262
2101 Constitution Avenue, NW
Washington, DC 2 0418
Telephone: 202/334-3129, FAX: 202/334-3718
e-mail: bscarbor@nas.edu

DELPHI SURVEY:
MANUFACTURING GRAND CHALLENGES IN THE YEAR 2020

QUESTIONNAIRE #1

Instructions

The objective of this survey is to provide a vision of the future manufacturing enterprise with its challenges and needs. ***Original thought and insight is required*** *to produce a vision that is more than a rehash of what has already been said. We know what the experts are saying about the trends in manufacturing today as they look to the future. What we want to know in this survey is what the experts will be saying as they look ahead to the year 2020 and beyond. .* ***Please extend your thinking beyond today's conventional wisdom.*** *Remember that the ground rules for manufacturing in 2020 will be much different than they are today.*

The intent is to define the ***profound changes*** *which will occur in manufacturing and* ***not the next incremental steps.***

Project yourself into the world of 2020 and beyond *and define what it will be and what the consequences will be for the manufacturing enterprise. You are not limited to predicting what will happen but also include what you think we should try to make happen.*

Try to spend approximately equal time on each question. Trial results for this survey indicated more extensive and thoughtful input on the first two questions. Insight and creative ideas are required on all four questions.

Your thinking should be expansive. However, for each question limit your input to the 3 to 5 most important ideas. Use as many sentences as you need to describe each idea.

For e-mail reply, please enter reply, then scroll down to the space after each question where you can insert your reply.

Questions

1. COMPETITIVE ENVIRONMENT
   The COMPETITIVE ENVIRONMENT for manufacturing will be dramatically different in the year 2020 and beyond. Major changes will occur in a number of different areas such as economics (national and global), education, competition, customers, geopolitics, ecological considerations, technology breakthroughs, relationships and agreements among nations, social conditions, and the workforce. For example, future cities look radically different, and all products are made from recyclable materials.

   Describe your vision of what this environment will be for the manufacturing enterprise in the year 2020 and beyond, by describing the dramatic and significant changes and events that will have occurred by then. Please be specific. Please do not just extrapolate from current trends.

2. ENTERPRISE
   Describe your vision of what the manufacturing ENTERPRISE will look like in the year 2020 and beyond.

3. CHALLENGES
   For the vision of the manufacturing enterprise that you provided for the year 2020 and beyond, what are the CHALLENGES that must be met?

4. TECHNOLOGY
   In order to meet these challenges, what are the major TECHNOLOGY developments that are needed? Technology is defined broadly to include resources, hardware, software, people finances, products, processing equipment, work processes, work designs, and business processes, etc.

(End of Survey)

NATIONAL RESEARCH COUNCIL
COMMISSION ON ENGINEERING AND TECHNICAL SYSTEMS
BOARD ON MANUFACTURING AND ENGINEERING DESIGN

COMMITTEE ON VISIONARY MANUFACTURING CHALLENGES

DELPHI SURVEY QUESTIONNAIRE #2
MANUFACTURING GRAND CHALLENGES IN THE YEAR 2020

Please return via email or fax by May 30, 1997 to:

Attn: Bonnie Scarborough
Board on Manufacturing and Engineering Design
Harris Building, Room 262
2001 Wisconsin Avenue, N.W.
Washington, DC 20007 USA
Phone: (202) 334-3562
Fax: (202) 334-3718
Email: bscarbor@nas.edu

NATIONAL RESEARCH COUNCIL
Board on Manufacturing and Engineering Design
2101 Constitution Avenue, NW
Washington, DC 20418

May 16, 1997

Dear Survey Participant:

I want to thank you for your contributions to the first round of this survey. We had worldwide participation in the first round with responses from over 180 manufacturing experts from 20 countries in Asia, Europe, North and South America, Australia, Africa, and the Middle East. Participants represented numerous industries, academia, and trade associations. The results reflected that a great deal of thought was put into the answers and we received many interesting ideas on the future for manufacturing.

This round will bring the survey to conclusion. In the attached questionnaire, we have summarized the prevalent answers given in the first round for the challenges facing manufacturing in 2020 and the enabling technologies. The objective of this round is to have you select which of these challenges and technologies are the most important for success in 2020 and to identify research areas for developing the priority technologies.

The results of this survey are intended to serve as a basis for establishing research agendas to meet the future needs of the manufacturing community. Therefore, it is important that you give thoughtful consideration to defining specific research areas and specific research topics that should be pursued. Please return the completed questionnaire by **May 30, 1997**. Thanks again for your efforts and we look forward to receiving your insights.

Sincerely,

John G. Bollinger
Chair, Committee on Visionary Manufacturing Challenges
Dean, College of Engineering
University of Wisconsin - Madison

DELPHI QUESTIONNAIRE #2

QUESTION 1: In the first round of the survey, respondents identified the following as major challenges for the manufacturing enterprise in 2020. Which of these challenges do you think are the most important? Place an "x" next to the **three** challenges that you consider most important for the manufacturing enterprise to succeed in 2020.

A. __ Enhancement of workforce performance and satisfaction to address rapidly changing and complex operational requirements, and diverse culture-based issues

B. __ Constant and concurrent development of innovative products, processes, and systems to meet shorter product life cycles, enhance value added, and advance manufacturing capabilities

C. __ Ability to rapidly develop and execute complex and dynamic alliances and collaborations

D. __ Response to severe environmental impact constraints and increasing material and energy scarcity

E. __ Achievement of the speed and flexibility to cost-effectively meet the ever-increasing customer demands for instant satisfaction with customized products

F. __ Adoption of rapidly developing technologies to increase and/or adapt the core strength of the enterprise to the marketplace

G. __ Development of an effective global infrastructure to support new optimal scale manufacturing configurations

H. __ Much more effective conversion of information to useful knowledge in an atmosphere of exploding availability of information

QUESTION 2: The following technologies were identified in round 1 of the survey as enablers for success of the manufacturing enterprise in 2020. Which technologies do you think are most important for enabling the manufacturing enterprise to meet its future challenges? Place an "x" next to the **five** technologies that you consider most important. For each of the technologies listed, descriptive information is enclosed in parentheses to further explain the technology; these descriptions are not meant to be inclusive or limiting.

A. __ Adaptable and reconfigurable manufacturing processes and systems (e.g., intelligent, mass customization, rapid creation of new production facilities, ability to accommodate wide range of product characteristics)

B. __ Systems model for all manufacturing enterprise operations (e.g., real time synthesis of planning, market demand, product development, distribution, social systems, wealth creation into manufacturing system planning, effective modeling of supply chains)

C. __ Micro and nano technology for fabrication processes (e.g., atom by atom fabrication of assemblies, development of microscale machines)

D. __ Processes to customize totally new materials with order of magnitude property improvements designed on the atomic scale (e.g., 10x strength improvement, defect-free materials, smart materials that can change properties in service in response to changing conditions, materials designed to be reprocessed or reconstructed)

E. __ Direct machine/user interfaces that enhance human performance and promote intelligent input (e.g., skill-leveraging, human commands transmitted directly to machine, human access to data via "bionic ears")

F. __ Net shape, programmable, flexible forming processes that require no hard tooling (e.g., pulsed power autoshaping, forming finished assemblies from the melt)

G. __ Design methods and manufacturing processes for products that can easily be reconfigured with software or hardware (e.g., products easily upgradable in the field for long life, products that can be customized by the customer)

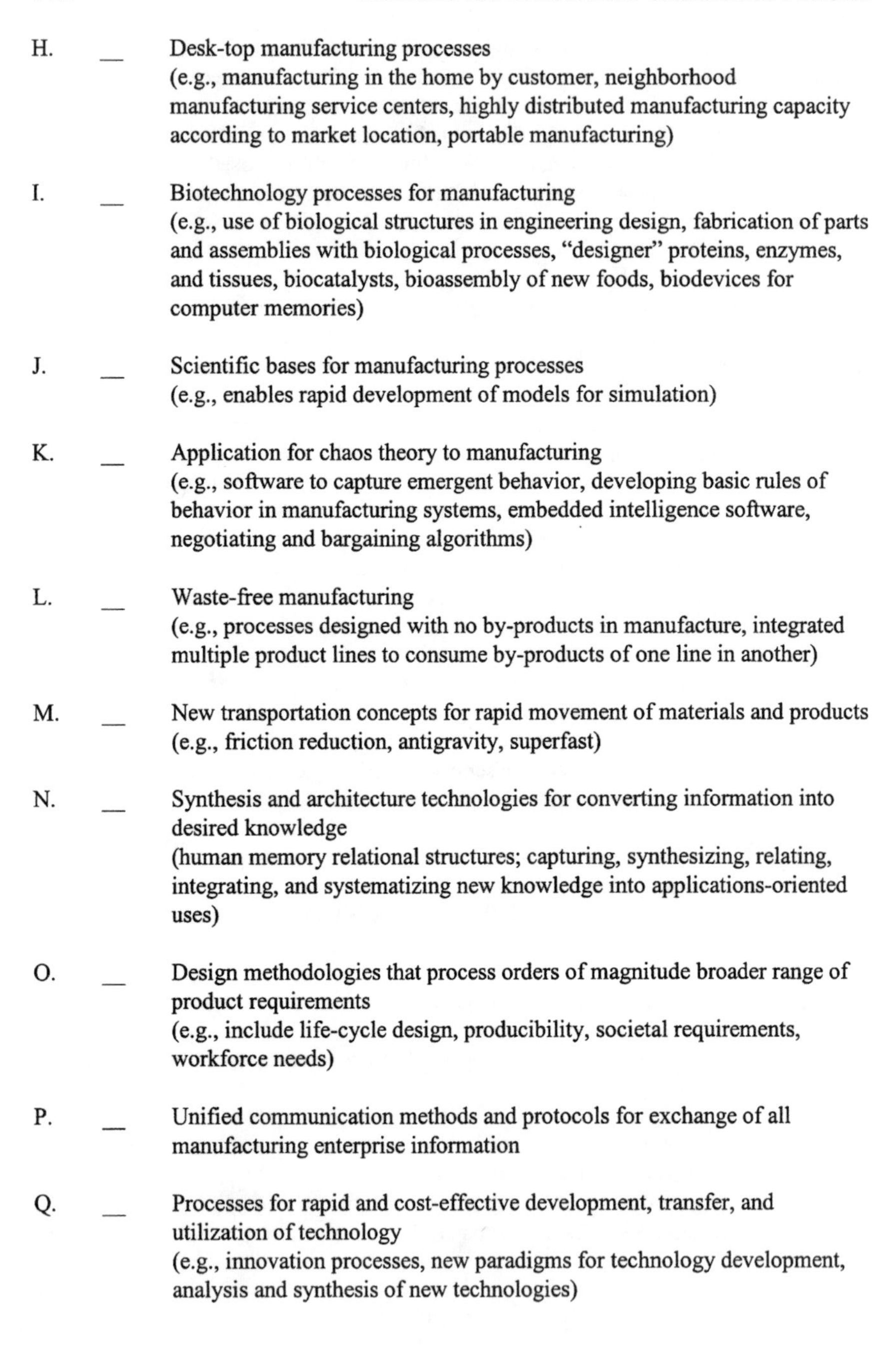

H. ___ Desk-top manufacturing processes
(e.g., manufacturing in the home by customer, neighborhood manufacturing service centers, highly distributed manufacturing capacity according to market location, portable manufacturing)

I. ___ Biotechnology processes for manufacturing
(e.g., use of biological structures in engineering design, fabrication of parts and assemblies with biological processes, "designer" proteins, enzymes, and tissues, biocatalysts, bioassembly of new foods, biodevices for computer memories)

J. ___ Scientific bases for manufacturing processes
(e.g., enables rapid development of models for simulation)

K. ___ Application for chaos theory to manufacturing
(e.g., software to capture emergent behavior, developing basic rules of behavior in manufacturing systems, embedded intelligence software, negotiating and bargaining algorithms)

L. ___ Waste-free manufacturing
(e.g., processes designed with no by-products in manufacture, integrated multiple product lines to consume by-products of one line in another)

M. ___ New transportation concepts for rapid movement of materials and products
(e.g., friction reduction, antigravity, superfast)

N. ___ Synthesis and architecture technologies for converting information into desired knowledge
(human memory relational structures; capturing, synthesizing, relating, integrating, and systematizing new knowledge into applications-oriented uses)

O. ___ Design methodologies that process orders of magnitude broader range of product requirements
(e.g., include life-cycle design, producibility, societal requirements, workforce needs)

P. ___ Unified communication methods and protocols for exchange of all manufacturing enterprise information

Q. ___ Processes for rapid and cost-effective development, transfer, and utilization of technology
(e.g., innovation processes, new paradigms for technology development, analysis and synthesis of new technologies)

R. __ Methodology for quantum jump in product and process reliability (e.g., variability reduction, new methods for robust design)

S. __ Low energy consumption processes (e.g., low-inertia machines, catalyst, alternate energy sources, high energy-density batteries)

T. __ New sensor technology for precision process control (real-time sensors for machine self-calibration, self-verification, self-correction, self-improvement)

U. __ 360 degree collaboration software (e.g., translate neural knowledge base to language that is personalized to different thinking styles; enable workforce participation in technology design and development; interactive visualization)

V. __ Low gravity, high vacuum manufacturing (practical manufacturing in space; earth-bound manufacturing facilities with space environment)

W. __ New educational methods (in-home facilities; smart and knowledge pills)

X. __ New concept manufacturing processes (e.g., ion beam, three dimensional chemical etching)

Y. __ New software design methods (e.g., robust, seamless, adaptive, inter-operable, highly reliable)

QUESTION 3: What research topics must be addressed to develop these technologies? For each of the five technologies that you identified as most important in Question 2, identify one or more specific research topics that must be addressed to develop that technology.

A major result of this survey is the research topics that you identify below. Please be as specific as you can so that there is sufficient definition to set research agendas. Your inputs should be phrased as topical statements with explanations in parentheses if needed. Explanations are encouraged where appropriate. Examples of research topics include: unifying theories leading to models for product producibility; new concepts and models for partitioning manufacturing systems; theories and defining experiments for human and machine or manufacturing systems interactions; processes for synthesizing biological structures; material design from first principles; concepts for nanofabrication machines.

Top 5 Technologies (identified by letter)	Research Topics
1.	
2.	
3.	
4.	
5.	

QUESTION 4: Which of the challenges facing manufacturing will be met by these technologies? For each of the five technologies that you identified as most important in Question 2, please indicate the challenges listed in Question 1 that will be met through the use of the technology. The challenges that you list here can be different from the top three challenges that you identified in Question 1.

Top 5 Technologies (identified by letter)	Challenges (identified by letter)
1.	
2.	
3.	
4.	
5.	

END OF SURVEY
Thank you for your participation.

Please return via email or fax by May 30, 1997 to:

Attn: Bonnie Scarborough
Board on Manufacturing and Engineering Design
Harris Building, Room 262
2001 Wisconsin Avenue, N.W.
Washington, DC 20007 USA
Phone: (202) 334-3562
Fax: (202) 334-3718
Email: bscarbor@nas.edu

APPENDIX D

Biographical Sketches of Committee Members

John G. Bollinger (chair) is dean of the College of Engineering and Bascom Professor of Engineering in the departments of electrical and computer engineering and industrial engineering at the University of Wisconsin at Madison. His 40-year career has been focused on computer control of machines and processes, robotics, design of production machinery, and analysis of dynamic systems. Dr. Bollinger has been actively involved with the manufacturing community as a consultant; as a member of the boards of directors of several companies in the consumer products, scientific instruments, machinery, and communications equipment industries; and as chairman of the board of an electronic drive systems company. He is a member of the National Academy of Engineering and the Board on Manufacturing and Engineering Design.

Dennis K. Benson is president of Appropriate Solutions, Inc., a public policy research and evaluation company founded in 1978. Dr. Benson has been involved in lifelong learning and development, with particular emphasis on workforce training and development, welfare reform, performance accountability, and quality management and enhancement systems. He is currently serving on the board of the National Association of Workforce Development Professionals and is champion for their professional development program.

Nathan Cloud is a professional engineer and president/founder of Cirrus Engineering, which offers consulting and design services for manufacturing enterprises. He recently retired from DuPont as engineering fellow. Over a period of 35 years, he has conceived and led development of innovative products and processes and advanced manufacturing systems and is named as the inventor on a number of U.S. patents. Mr. Cloud helped DuPont incorporate advanced manufacturing technology into its operations in a variety of ways, including leading their Advanced Manufacturing Program and Laboratory and a number of Next Generation Manufacturing project initiatives. He also led the development of new business processes for the concurrent creation and development of products and their manufacturing systems, one of which was adopted throughout DuPont as a "best practice" for new product development. Mr. Cloud is continuing to develop software that supports the creation of new business enterprises.

Gordon Forward is vice chairman of the board of TXI. Prior to assuming his current position, Dr. Forward was president and chief executive officer of Chaparral Steel. His experience is in business management, application of advanced technology in manufacturing enterprises, and environmentally conscious manufacturing. Dr. Forward is the chair of the Business Council for Sustainable Development, Gulf of Mexico Chapter, an organization of international business people seeking solutions to environmental problems. He is a director of the Steel Manufacturing Association and a director of Novanda Forest, Inc. Dr. Forward is a member of the National Academy of Engineering.

Barbara Fossum is director of the Master of Science Degree Program in Science and Technology Commercialization at the IC2 Institute, co-director of the Manufacturing System Center, and member of the teaching faculty at the University of Texas. She has been on the Advisory Board of the University of Texas Quality Center and was previously director and founder of the Quality Management Consortia Program at the university. Dr. Fossum has expertise in operations management, computer systems for manufacturing, total quality management practices, business process reengineering, and small company operations. She is a fellow of the Society of Manufacturing Engineers (SME) and is on their International Board of Directors. From 1991 to 1995, she served on the Board of Advisors of the Computer and Automated Systems Association of SME. Dr. Fossum has been involved in a number of manufacturing initiatives, including the Next Generation Manufacturing project, National Industrial Competitiveness Workshops for Information Systems in Manufacturing, and Computer Aided Acquisition and Logistics Support and Concurrent Engineering.

Donald Frey is professor of industrial engineering and management sciences at Northwestern University. Prior to joining Northwestern in 1988, he held positions with the Ford Motor Company and Bell and Howell, where he was chairman of the board and chief operating officer. At Ford, he held positions in research, product development, and operations, including vice president for product development and general manager of the Ford Division. Dr. Frey has experience in enterprise management, strategic planning, and technology research and development. He has served on the boards of directors for several multinational companies and has worked with the World Bank on economic development issues. Dr. Frey is a member of the National Academy of Engineering.

David Hagen is immediate past president of the Engineering Society of Detroit. Previously, he was president of the Michigan Center for High Technology, which is responsible for implementation of technological innovations into industrial applications. Mr. Hagen spent 35 years at Ford Motor company, where he was Chief Engine Engineer; general manager of the Engine Division, responsible for eight manufacturing plants and engine design engineering; and general manager of the ALPHA program, responsible for developing advanced processes in product design and manufacturing operations for worldwide application. Mr. Hagen has experience in product design, manufacturing operations, and implementation of advanced design and processing technologies.

James A. Jordan, Jr. is president and cofounder of NGM Knowledge Systems. He organized the Next Generation Manufacturing System's Intelligent Manufacturing Systems Program and the Next Generation Manufacturing Education Program for the Consortium for Advanced Manufacturing-International. Dr. Jordan has cochaired the Agile Virtual Enterprise Executive Development Group and was lead author for the Next Generation Manufacturing project's Imperative on Enterprise Integration. He has been a contributor to several other national and international studies of the future of manufacturing. Dr. Jordan retired in 1993 after a lengthy management career with IBM focused on research and development for information technology and systems to support industry. He holds a Ph.D. in physics from the University of Michigan.

Ann Majchrzak is professor of information systems in the Department of Information and Operations Management at the Marshall School of Business of the University of Southern California. Her research interests are human factors and workforce issues related to advanced manufacturing, including agile manufacturing, application of artificial intelligence, concurrent engineering, implementation of advanced manufacturing technology, programmable manufacturing technology, and tools for interdisciplinary manufacturing systems. Professor Majchrzak has authored two books on human factors: *Human Aspects of Computer Aided Design* and *The Human Side of Factory Automation: Managerial and Human Resource Strategies for Making Automation Succeed.*

Eugene Meieran is an Intel fellow at Intel Corporation, which produces semiconductor chips and devices. He has been a leader in the development and implementation of world-class manufacturing initiatives at Intel. His expertise is in semiconductor materials and processes, electronic packaging, process control and statistics, and application of artificial intelligence in manufacturing. He has contributed to many national and international manufacturing initiatives, including the Next-Generation Manufacturing project, the Massachusetts Institute of Technology Leaders for Manufacturing program, and the National Research Council Committee on Information Technology for Manufacturing. Dr. Meieran is a member of the National Academy of Engineering.

David Miska is manager of United Technologies Corporation's Manufacturing Council, which is responsible for the introduction of new manufacturing technology and practices for all domestic and international operations. He is also a project manager for United Technology's supply chain management initiatives. Prior to his current assignment, Mr. Miska held a number of positions at the Pratt and Whitney division of United Technologies where he was responsible for machine and tool planning, manufacturing engineering and technology, quality assurance, CAD/CAM systems, customer support, and product engineering. Mr. Miska has served as chair of the American Society of Quality Control and industry chair for the production equipment and systems group at the National Center for Manufacturing Sciences.

Lawrence J. Rhoades is president and chief executive officer of Extrude Hone Corporation, a process developer and equipment supplier for a wide range of manufacturing industries. He holds patents on more than two dozen inventions related to nontraditional manufacturing processes for machining, finishing, forming, and measurement. He has been the chair for the advisory committee of the U.S. Export-Import Bank, and has served on numerous advisory groups for the U.S. Department of Defense and the U.S. Department of Commerce, addressing technologies and business practices related to manufacturing. Mr. Rhoades has served on the board of the Association for Manufacturing Technologies and currently serves on the boards of Concurrent Technologies Corporation, the National Center for Manufacturing Sciences, and the National Institute of Standards and Technology's National Manufacturing Extension Partnership program.

Eugene Wong is the chief scientist and member of the board of directors of Vision Software Tools, Inc. He was recently appointed to head the Engineering Directorate at the National Science Foundation. He was a member of the faculty at the University of California at Berkeley from 1962 to 1994, a founder of the INGRES Corporation, and associate director of the White House Office of Science and Technology Policy. His recent interests have been in software systems, and he is currently developing software products for automating business processes. Dr. Wong is a member and councilor of the National Academy of Engineering.

Index

D

E

G

H

I

K

L

M

N

O

P

U

W